"These two eminent astronomers, one from Australia and one from South Africa, bring a unique perspective to the faith and science arena. What they reveal about Galileo—who is often cited as an example of the great divide—demonstrates instead that strong faith and expert science can go together. Indeed, the authors themselves follow in Galileo's path, approaching both fields with a spirit of humility and wonder."

Philip Yancey, author, *What's So Amazing About Grace?* and *The Jesus I Never Knew*

"Galileo showed us how to write in the book of nature, but his world read only from the book of Scripture—thus descended a debate that tore Galileo's world apart and has never been reconciled, even to our time. *God and Galileo* is a personal journey through the world of two books, nature and Scripture, guided by leading astronomers who have wondered, like many others, why we cannot seem to read clearly from both books at the same time. Their conclusion is that we can and, to reach our fullest understanding, we should. Galileo concluded the same but was not allowed to speak it. *God and Galileo* finally gives him a voice."

Bruce Elmegreen, Astrophysicist, Thomas J. Watson Research Center, IBM; recipient, Dannie Heineman Prize for Astrophysics (2001)

"With so many scientists seeing Christian faith as irrelevant to scientific truth and so many Christians seeing science as contradictory to Christian truth, this unique, groundbreaking, and deeply researched book by two believing, distinguished, and top-drawer astronomers is one that had to be written. It makes clear that the totality of truth has to be drawn on the one hand from the book of Scripture, with its story of grace and incarnation, and on the other hand from the book of nature, with its story of space and matter. Both books are vital to the full comprehension of reality, and the authors show this with convincing clarity. We dare not be blind either to nature or Scripture, whose respective truths are complementary, not contradictory, because both have the same author. *God and Galileo* brings us unique perspectives and insights related to faith, grace, and astronomy not evident in any other contemporary writing. My prayer is that it will be a landmark contribution to this debate and a classic both for today and for generations to come."

Michael Cassidy, Founder, African Enterprise; Honorary Cochair, Lausanne Movement; author, *The Church Jesus Prayed For*

"*God and Galileo* needed to be written. The majority of scientists today are either atheist or agnostic, and there is rarely any discussion about the relationship between the physical and spiritual realms of knowledge. In scientific circles, these subjects mix like oil and water. Yet the relationship between a Creator and the origin of the universe is an important subject of fundamental interest to everyone. Is there a connection between science and religion, or are the two in conflict as completely independent realms of knowledge? This book addresses this question head-on. Written by two leading international researchers in astronomy, the book reflects extensively on the interaction between the universe of space and the God of grace. To make their point, the authors offer personal and contemporary reflections on a 1615 letter written by Galileo Galilei, in which he addresses this very conflict between revelation and reason. *God and Galileo* is a devastating attack on the dominance of atheism in science today. It is a must-read, offering proper perspective on life and why we exist in the universe."

Giovanni Fazio, Senior Physicist, Harvard-Smithsonian Center for Astrophysics; Fellow, American Physical Society; recipient, Henry Norris Russell Lectureship (2015)

"In a world growing increasingly hostile to Christianity, clarity is our first and best defense. Indeed, the challenge for the believer today is to tread fearfully in such a world and to remain true—that, and to be well informed. Among other things, that means exercising caution when choosing whom to listen to. This is one of the great payoffs of this book. *God and Galileo* is about clarity in its best and most attractive sense. Using the words of Galileo Galilei as a prop, and with language accessible to the general reader, astronomers Block and Freeman conduct an intimate dialogue with history. Tampering with deep cultural memory, they explore the harmonies and agreements that exist between the book of nature and the book of Scripture, being, as they were, according to Galileo, crafted by the same author."

David Teems, author, *Tyndale: The Man Who Gave God an English Voice*

GOD
AND GALILEO

GOD

AND GALILEO

What a 400-Year-Old Letter
Teaches Us about Faith and Science

David L. Block and Kenneth C. Freeman

WHEATON, ILLINOIS

God and Galileo: What a 400-Year-Old Letter Teaches Us about Faith and Science
Copyright © 2019 by David L. Block and Kenneth C. Freeman
Published by Crossway
    1300 Crescent Street
    Wheaton, Illinois 60187

All rights reserved. No part of this publication may be reproduced, stored in a retrieval system, or transmitted in any form by any means, electronic, mechanical, photocopy, recording, or otherwise, without the prior permission of the publisher, except as provided for by USA copyright law. Crossway® is a registered trademark in the United States of America.

Appearing in the appendix by permission of Oxford University Press is Galileo, *Letter to the Grand Duchess Christina*, trans. Mark Davie, in *Galileo: Selected Writings*, trans. William R. Shea and Mark Davie, Oxford World's Classics (New York: Oxford University Press, 2012), 61–94. Citations are omitted for quotations from Galileo's letter that appear in this book, and when such quotations appear as extracts, they are italicized.

Cover design: Derek Thornton, Faceout Studios

Cover images: *A Discourse concerning a New World & Another Planet in 2 Bookes*, 1640 (frontispiece), by William Marshall (fl. 1617–1649) / private collection / Bridgeman Images. *Galileo (and Urban VIII)*, 19th century, by Edmond Theodor van Hove (1853–1913) / Mondadori Portfolio / Bridgeman Images.

First printing 2019

Printed in the United States of America

Scripture quotations marked DRB are from the Douay-Rheims Bible.

Scripture quotations marked ESV are from the ESV® Bible (The Holy Bible, English Standard Version®), copyright © 2001 by Crossway, a publishing ministry of Good News Publishers. Used by permission. All rights reserved.

Scripture quotations marked GNB are from the *Good News Bible* © 1994 published by the Bible Societies / HarperCollins Publishers Ltd., *UK Good News Bible* © by American Bible Society 1966, 1971, 1976, 1992. Used with permission.

Scripture quotations marked KJV are from the *King James Version* of the Bible.

Scripture quotations marked MESSAGE are from *The Message*. Copyright © by Eugene H. Peterson 1993, 1994, 1995, 1996, 2000, 2001, 2002. Used by permission of NavPress Publishing Group.

Scripture references marked NKJV are from *The New King James Version*. Copyright © 1982, Thomas Nelson, Inc. Used by permission.

Scripture references marked TLB are from *The Living Bible* © 1971. Used by permission of Tyndale House Publishers, Inc., Wheaton, IL 60189. All rights reserved.

All emphases in Scripture quotations have been added by the author.

Hardcover ISBN: 978-1-4335-6289-1
Epub ISBN: 978-1-4335-6292-1
PDF ISBN: 978-1-4335-6290-7
Mobipocket ISBN: 978-1-4335-6291-4

Library of Congress Cataloging-in-Publication Data

Names: Block, David L., author. | Freeman, Ken, 1940– author.
Title: God and Galileo : what a 400-year-old letter teaches us about faith and science / David L. Block and Kenneth C. Freeman.
Description: Wheaton, Illinois : Crossway, 2019. | Includes bibliographical references and index.
Identifiers: LCCN 2018030455 (print) | LCCN 2018052878 (ebook) | ISBN 9781433562907 (pdf) | ISBN 9781433562914 (mobi) | ISBN 9781433562921 (epub) | ISBN 9781433562891 (hc) | ISBN 9781433562921 (ePub) | ISBN 9781433562914 (Mobipocket)
Subjects: LCSH: Religion and science. | Faith and reason—Christianity. | Astronomy—Religious aspects—Christianity. | Galilei, Galileo, 1564–1642—Influence.
Classification: LCC BL240.3 (ebook) | LCC BL240.3 .B6245 2019 (print) | DDC 261.5/5—dc23
LC record available at https://lccn.loc.gov/2018030455

Crossway is a publishing ministry of Good News Publishers.

LB		28		27		26		25		24		23		22		21		20		19
15	14	13	12	11	10	9	8	7	6	5	4	3	2	1						

We dedicate our book to the memory of the martyr William Tyndale. His translations have been pivotal to our readings of the New Testament and to the grace of God described therein. As W. R. Cooper noted, "The printing in 1526 of William Tyndale's translation of the New Testament from Greek into English was arguably the most important single event in the history of the English Reformation."

Contents

List of Illustrations 11

Preface 13

Acknowledgments 17

PART 1
GRACE AND SPACE

Setting the Stage 23

1 Is There Grace in Space? 27

2 Misunderstanding Truth 39

3 Understanding the Universe and Scripture 49

4 What Grace and Space Cannot Tell Us 63

5 The Fraud of Scientism 71

6 An Illusion of Conflict 93

7 Discerning the Truth 109

8 The Two Cathedrals 121

PART 2
HISTORICAL VIGNETTES

9 A Moon of Glass from Murano, Venice 139

10 A Troubled Dinner in Tuscany 149

11 Winning Back Trust: Astronomy and the Vatican 155

PART 3
PERSONAL EXPERIENCES OF GRACE

12 Grace in the Life of Blaise Pascal 163
13 Grace alongside a Telescope in South Africa 167

Appendix: Galileo's *Letter to the Grand Duchess Christina of Tuscany*..179
Bibliography and Additional Readings 214
General Index 220
Scripture Index 224

Illustrations

Figures

1. Abell 2218 cluster of galaxies
2. The northern lights
3. Windblown trees in Twistleton Scar in the Yorkshire Dales
4. Title page to William Tyndale's 1526 translation of the New Testament
5. Professor Jean Mesnard
6. Composite of the planet Saturn with six of its moons
7. 360-degree panorama of the southern sky
8. Rosette Nebula
9. Spiral galaxy Messier 83
10. Jupiter and its four "Medicean" moons
11. The villa Il Gioiello, Galileo's last home
12. Andromeda spiral galaxy
13. Cosmologist Allan Sandage
14. Ice glass
15. *Portrait of Galileo Galilei,* by Giacomo Ciesa, 1772–1773

Photographs follow page 160.

Preface

The truth of nature belongs to the physical or scientific realm. In contrast, the much broader nature of truth includes both the physical and spiritual domains; God's revelation of himself to us is the work of his grace. To insist that truth lies in only one or the other domain is only half the story, as in watching trees swaying and bending without recognizing the presence of the wind.

In earlier times, the church and its cardinals ruled supreme and misused the book of Scripture, claiming that it asserted things about science that it did not assert, all the while paying little regard for experimental science. This situation was clearly out of balance. Scripture is wholly true in all that it claims, and when interpreted rightly, it harmonizes perfectly with the book of nature. The church had misunderstood this principle and used Scripture to silence science.

The situation today is equally out of balance, to the other extreme. The scientific book of nature is paramount today, and many high-profile scientists would have us abandon the Scriptures entirely as a source of truth about our world. The philosophical viewpoint of these self-appointed "cardinals of science" is driven as much by the mood of the age and the personalities and beliefs of individuals as it is by scientific data and rigorous theory.

Today atheist fundamentalism rules, with its basic philosophical agenda to avoid any need for a Creator. In this book, we, as two professional astronomers, reflect on the universe of space and the grace of God. We comment on the subjective and territorial nature of science and affirm that the science of today is not in a position to pronounce on the existence of God. We argue that God is ultimately known not through human logic or experiment but through his self-revelation.

Our reflections sweep from the universe of galaxies to the universe of the heart. In the words of Blaise Pascal, "The heart has its reasons, which reason does not know. We know the truth not only by the reason, but also by the heart."[1] This is God's universe, wherein grace prevails: we need to be receptive to both reason and revelation. It should not surprise us that people who are trying to make sense of this world are provided with a map by the Maker of this world, who, by his grace, has visited his world in person.

It is these shared beliefs that led to the writing of this book. Our story starts at the Siding Spring Observatory near Coonabarabran, some 350 miles north of Canberra and home to the largest optical telescope in Australia. We had completed our work at the observatory, and we started driving back to Canberra. Something was on our minds while we were on our long drive that day. We had been thinking about writing a book on science and our personal relationship with God. While many books have been written about the harmony of science and God, we had encountered several books in Australia by scientists that have swayed the opinions of many away from God. But there was a *new* perspective we could add.

From Coonabarabran we drove through the town of Parkes, in Western New South Wales. Parkes is another astronomy town, with its huge radio telescope looming on the horizon. After a few more hours of talking in the car, we reached Canberra, and our book had been conceived. We were determined not to add yet another book to the God-science debate but rather to share our thoughts on the grace of God in the context of science. To be specific, we aimed to consider the grace of God, who had entered the restrictions of space and of time. Pascal spoke of our God humbled, our God weeping, our God dying, Jesus incarnate. In more modern times, Albert Einstein, one of the greatest scientists of all time, refers to Jesus as "the luminous figure of the Nazarene."[2]

---

1. From Blaise Pascal, *Pensées* (1670), in Blaise Pascal, *Thoughts*, trans. W. F. Trotter, in vol. 48 of *The Harvard Classics*, ed. Charles W. Eliot (New York: Collier, 1910), 99. The original French quote reads thus: "Le cœur a ses raisons, que la raison ne connaît point." Blaise Pascal, *Pensées, Fragments et Lettres de Blaise Pascal*, ed. Armand-Prosper Faugère (Paris: Andrieux, 1844), 2:172.

2. "What Life Means to Einstein: An Interview by George Sylvester Viereck," *Saturday Evening Post*, October 26, 1929, 117, http://www.saturdayeveningpost.com/wp-content/uploads/satevepost/einstein.pdf.

Our book took us to Florence and Venice, key places in Galileo Galilei's life; to the secret Vatican archives in Rome; to the island of Murano, famed worldwide for the production of glass, a crucial component in the manufacture of many optical telescopes; and to archives where we held the original manuscripts of Galileo and Nicolaus Copernicus in our hands.

It also took us to France, the land of Blaise Pascal (1623–1662). The genius of Pascal as a scientist is beyond question. But what has intrigued us about Pascal was his experience with God, of knowing God—his "night of fire," describing his cataclysmic personal encounter with God in 1654. We wished to appreciate the greatness of the scientist Pascal—to enter sympathetically into the spirit of the age in which he lived—and for that we had many discussions in Paris with the late Professor Jean Mesnard, one of the greatest experts of our time on Pascal. Pascal never came away with religion but with the wonder of actually knowing God, at a personal level. He penned his innermost feelings, which became his famous *Les Pensées*.

Here we offer our thoughts and experiences, as astronomical researchers with careers jointly spanning more than ninety-five years. They are our thoughts. Others may disagree. Are we trained theologians? No, but then neither were Peter the fisherman or Matthew the tax collector. Even as we benefit greatly from trained Christian theologians, grace is freely imparted to all willing to receive.

Our first theme focuses on the modus operandi of scientists, their methodologies, their goals, the manner in which science has largely molded the prevailing mood of our age, and the limitations of the scientific method. These are insights that people who are not scientists may be unaware of. Our second theme is a question: When did science lose its grace? Here is what we mean: When and why have so many scientists today been blinded to the grace of God, unlike scientific giants like Pascal and Galileo?

Galileo makes a particularly constructive contribution to these questions in his famous 1615 *Letter to the Grand Duchess Christina of Tuscany*. His letter is full of insights into the science-religion interface, and we use it as a basis for our book. This is not a book about Galileo. However, excerpts from the letter appear throughout our discussion in italics, and we use the letter as a springboard for our perspectives as professional astronomers, four hundred years later, on the mood of our age and our own exposure and response to God's grace and revelation. In Galileo's time, theology was the queen of all the sciences. Today, in the minds of multitudes, science is god. Is science indeed the new savior of mankind?

Readers of the atheist literature may come away from such volumes believing that science has made God unnecessary. We wrote this book to offer a different perspective. Although science can illuminate the glories of the creation, we argue that it is beyond the domain of science to infer that God does not exist. God exists outside space and time. Science does not have the weapons to expunge God's Spirit or the revelation of his spiritual kingdom. At the heart of God's kingdom is grace.

This is a book about the God whom we have come to know—both through Scripture and in personal experience. In a universe spanning approximately 92 billion light years, we have come to know its Creator.

Acknowledgments

We are deeply grateful to Mark Davie, Honorary University Fellow at the University of Exeter, for his willingness to share his expert, modern, and lively translation of Galileo's letter from Italian with us. We also thank Oxford University Press for permission to reproduce Galileo's letter in full at the end of our book.

Our thanks also go to past directors of the Vatican Observatory, George Coyne for insights into Galileo's *Letter to the Grand Duchess* and José Funes for facilitating our visit to the Vatican. It is a pleasure to thank Massimo Bucciantini (University of Siena), Michele Camerota (University of Cagliari), Franco Giudice (University of Bergamo), Mario Biagioli (University of California Davis), and Daniele and Ottavio Besomi of Switzerland for their insights.

Our first visit to Florence was organized by the late Francesco Palla, director of the Arcetri Observatory, who had been introduced to us by our mutual friend and colleague Joe Silk (Emeritus Savilian Professor at Oxford University). We are indebted to Palla for showing us Galileo's final home, Il Gioiello, and for contacting Isabella Truci and Susanna Pelle at the Biblioteca Nazionale Centrale di Firenze—they graciously allowed us to examine manuscripts written in Galileo's hand. Palla also introduced us to the optician Giuseppe Molesini, of the Istituto Nazionale di Ottica in Florence, and we thank him for helpful discussions concerning Galileo's telescopes and the optical testing of his lenses. The hospitality extended by Edvige Corbelli in Florence to one of us (David) when visiting Il Gioiello a second time is gratefully acknowledged.

We remain indebted to the legendary Venetian historian on glass Rosa Barovier Mentasti for the time that she spent with one of us

(David). Galileo had compared the appearance of the moon through spyglasses or telescopes to that of cracked or "ice glass." We were fascinated by this comparison and are very grateful to Rosa Barovier Mentasti for explaining this to us. No talk of the history of glass in the Veneto would be complete without mentioning the Barovier family, predating the time when Galileo walked the streets of Venice.

In Rome, we were privileged to have a discussion with Giuseppe Tanzella-Nitti of the Pontificia Università della Santa Croce, focusing on the subject of the book of nature and the book of Scripture. We also thank Marinella Calisi for allowing us to examine a published copy of Galileo's *Sidereus Nuncius* at the Osservatorio Astronomico di Roma.

We are deeply grateful to the late Jean Mesnard, who was one of the great experts on Blaise Pascal. Mesnard was emeritus professor of the Sorbonne, and we had wide-ranging discussions with him in Paris over several years. These discussions centered on the thoughts of Pascal but also engaged the thoughts of Galileo and René Descartes. We express our profound appreciation to Sarah Frewen-Lord of Ravi Zacharias International Ministries for transcribing our discussions with Professor Mesnard in Paris and translating them from French into English.

We warmly thank Walter C. Kaiser Jr., president emeritus and distinguished professor of Old Testament at Gordon-Conwell Theological Seminary in Massachusetts, for valuable insight into biblical exegesis.

We enjoyed visits to Owen Gingerich, professor emeritus of astronomy and the history of science at Harvard University, for discussion—and for the privilege of examining various precious historical books, including an early edition of *De revolutionibus orbium coelestium* by Nicolaus Copernicus. We thank him and our colleague George Ellis, emeritus distinguished professor of complex systems in the Department of Mathematics and Applied Mathematics at the University of Cape Town in South Africa, for correspondence and for sending us some of their articles.

Martin Rees, Astronomer Royal, sent us his tribute to Stephen Hawking. For this and for his continuing support, we are very grateful.

Our visit to Poland was greatly enriched by Professor Staszek Zola, who assisted us in matters historical and gave us the opportunity to examine an original manuscript by Nicolaus Copernicus. Of much assistance to us in Poland was the librarian Dorota Antosiewicz; she spared no efforts in facilitating the examination of rare astronomical treatises. The funding for our visit to Poland was enabled by the persistent efforts of Louise Hirsch in Melbourne, to whom we are very grateful; our thanks go to Rita Ivenskis, Linda Bedford, Ros and Hedy Garson, Margaret Mascara, and Louise and Lawrence Hirsch.

We extend our deepest gratitude to Anna Teicher in Cambridge for so readily translating a section (from Italian into English) of a poem by Galileo for us, titled "Against Wearing the Gown." We also thank the director, Professor Paolo Galluzzi, and librarian, Daniela Pozzi, at the Museo Galileo in Florence for their assistance. We profoundly thank Sara Bonechi at the Museo Galileo for her deep insights into Galileo's poem and for introducing us to Anna Teicher.

We acknowledge with gratitude the advice of John Mather (Senior Project Scientist for the James Webb Space Telescope) and Charles Bennett (Johns Hopkins University) regarding the interpretation of maps of our very early universe.

Our warm thanks go to Philip Yancey, David Teems, Giovanni Fazio, Bruce Elmegreen, and Michael Cassidy for writing endorsements of our book. We are also grateful to David Munro, Mark and Heather Stonestreet, and Anke Arentsen for reading a first draft of our book and for offering their inspiring thoughts. We are indebted to Celeste Johnson for her fastidious accuracy in proofreading Galileo's letter in the appendix.

It is a singular joy to thank Peter Loose for his immense support and enthusiasm for our project. Loose's passion for *God and Galileo* has been a source of great encouragement to us. It was Loose who introduced us to Dave DeWit, vice president of book publishing at Crossway. Communicating with Dave DeWit has shown both of us the meaning of the word *grace* in the truest sense of the word; a more gracious set of communications no author could ever hope for.

It is a great pleasure to thank our editor at Crossway, David Barshinger, for his meticulous attention to detail combined with his profound professional insights into the structuring of our text. His editorial skills are astonishing. His firm and pleasant interaction with us throughout the editing process is greatly appreciated.

We warmly thank Samuel James at Crossway for so ably assisting us to interweave the letter by Galileo into an orderly flow in our book. And we are grateful to Jill Carter for all her administrative assistance.

It is a joy to thank Andreas Kahlau at Silvertone International and Gert Rautenbach at the RGB Pixellab for their photographic expertise. Profound appreciation is expressed to Udo Kieslich at the College of Digital Photography in Johannesburg for his professional insights and assistance.

We extend our deepest appreciation and heartfelt thanks to our wives, Liz Block and Margaret Freeman, for their encouragement and exceptional patience over a good couple of years while we were working on this book; they held the fort at home while we traveled to different countries.

Finally, we warmly thank our friend Michal Dabrowski. The tone of our book was changed by his emails. It was Michal who urged us not to write simply a historical account of Galileo's scientific and religious challenges of long ago but to also focus on the grace and revelation of God in our personal walk with him.

To each person who helped us synthesize our thoughts on the universe of space and the God of grace, we thank you.

David L. Block and Kenneth C. Freeman
Johannesburg and Canberra
2018

PART 1

GRACE AND SPACE

Setting the Stage

The dispute between the church and Galileo sowed the seed for the apparent divorce between science and faith. The dispute was about the theory of the universe, presented by Nicolaus Copernicus (1473–1543) in 1543, that the sun was at the center of the universe. This theory was in opposition to the Aristotelian view promoted by the church, that the sun and other planets were in orbit around the earth. Galileo favored the Copernican model because of what he observed through his own telescopes, particularly that the moons of Jupiter were in orbit around the planet Jupiter. These were landmark telescopic observations—not all bodies in the universe were orbiting the earth!

Copernicus's theory was regarded as heretical because it clashed with the church's interpretation of the biblical creation account, in which God "set the earth on its foundations" (Ps. 104:5 ESV). Harvard historian Owen Gingerich carefully elaborates:

> As far as the theologians were concerned, the Copernican system was not really the issue. I can hardly emphasize this point enough. The battleground was *the method* itself, *the route* to sure knowledge of the world, *the question* of whether the Book of Nature could in any way rival the inerrant Book of Scripture as an avenue to truth.[1]

Who controls the access to the wells of truth?

Pope Urban VIII allowed Galileo to continue his investigations of the heavens, provided his findings were presented as theory, not

---

1. Owen Gingerich, "The Galileo Affair," *Scientific American* 247, no. 2 (1982): 123–24; italics added.

as fact. But in the end, Galileo could not restrain himself from fully embracing the heliocentric system.

Galileo was summoned from Florence to Rome for trial by the Inquisition in 1633. He saw no conflict between the domains of scientific research and faith in God. He believed that study of the universe would promote greater understanding of the correct interpretation of the Scriptures. But the label of Galileo as a suspected heretic prevailed in the trial, and he was forced to recant and sentenced to house arrest: he died in Arcetri and on January 9, 1642, was buried in an unmarked grave.[2]

Galileo was far-reaching in his views: he saw the difference between the nature of truth (Scripture) and the truth of nature (science).[3] Although it soon became clear that Galileo's worldview was correct, it took until 1992 for the church to offer an acknowledgment of the error of the theologians at the time. Here are the words from Pope John Paul II:

> Thanks to his intuition as a brilliant physicist and by relying on different arguments, Galileo, who practically invented the experimental method, understood why only the sun could function as the centre of the world, as it was then known, that is to say, as a planetary system. The error of the theologians of the time, when they maintained the centrality of the earth, was to think that our understanding of the physical world's structure was, in some way, imposed by the literal sense of Sacred Scripture. Let us recall the celebrated saying attributed to Baronius: "*Spiritui Sancto mentem fuisse nos docere quomodo ad coelum eatur, non quomodo coelum gradiatur*" ["It was the Holy Spirit's intent to teach us how one goes to heaven, not how the heavens go"]. In fact, the Bible does not concern itself with the details of the physical world, the understanding of which is the competence of human experience and reasoning. There exist two realms of knowledge, one which has its source in Revelation and one which reason can discover by its own power. To the latter belong especially the experimental sciences and philosophy. The distinction between the two realms

---

2. Maurice A. Finocchiaro, ed. and trans., *The Essential Galileo* (Indianapolis: Hackett, 2008), 24. The present whereabouts of Galileo's mortal remains will be discussed later.
3. This eloquent expression comes from Gingerich, "Galileo Affair," 119.

of knowledge ought not to be understood as opposition. The two realms are not altogether foreign to each other, they have points of contact. The methodologies proper to each make it possible to bring out different aspects of reality. . . .

Thus the new science, with its methods and the freedom of research which they implied, obliged theologians to examine their own criteria of scriptural interpretation. Most of them did not know how to do so.

Paradoxically, Galileo, a sincere believer, showed himself to be more perceptive in this regard than the theologians who opposed him.[4]

The pope also refers in his address to Galileo's famous letter dedicated to Christina of Lorraine (1565–1637), the Grand Duchess of Tuscany. Christina of Lorraine was the favorite granddaughter of Catherine de Medici, the queen of France, and Christina's son Cosimo II de Medici (1590–1621) was Galileo's patron. Dedicating this letter to the Grand Duchess Christina was a very prudent move by Galileo, as discussed below. It has even been suggested that while Galileo's letter of 1615 to the Grand Duchess was indeed dedicated to her, it was never intended to be read by her. In fact, there are no records that the Grand Duchess actually read the letter.[5]

As Pope John Paul II emphasized, it was a battle for the soul of the world then, and it is a battle for the soul of the world now. What better aid for us to use four centuries later than the actual letter written by Galileo in 1615 and addressed to the Grand Duchess of Tuscany? The letter is about the harmony between the new science and faith in God. It is a letter of such significance that it does not escape mention by Pope John Paul II.

---

4. John Paul II, "Allocution of the Holy Father John Paul II to the Participants in the Plenary Session of the Pontifical Academy of Sciences," October 31, 1992, in *Papal Addresses to the Pontifical Academy of Sciences 1917–2002*, Pontificiae Academiae Scripta Varia, no. 100 (Vatican City: Pontificia Academia Scientiarum, 2003), 336–43.

5. We are indebted to Ottavio and Daniele Besomi, Massimo Bucciantini, Michele Camerota, and George Coyne for their insights into these aspects of the *Letter to the Grand Duchess Christina of Tuscany*.

1

Is There Grace in Space?

The Two Books

Galileo began his *Letter to the Grand Duchess Christina of Tuscany* as follows:

> *A few years ago, as your Highness well knows, I discovered many things in the heavens which had been invisible until this present age. Because of their novelty and because some consequences which follow from them contradict commonly held scientific views, these have provoked not a few professors in the schools against me, as if I had deliberately placed these objects in the sky to cause confusion in the natural sciences.*[1]

A recurring theme in this letter, and a source of great concern to Galileo, was this tension between what he observed through his telescope and the opinions of the theologians. Cherished by the theologians of the day was Aristotle's geocentric model of the universe, wherein all bodies, including the sun, orbited the earth. The earth was perceived to be the center of the universe. At the time of Galileo, the book of Scripture was used by many as the only source of truth, and the concept of a non-earth-centered world, as revealed by Nicolaus Copernicus's and Galileo's new observations, was seen as a huge threat.

The shoe is now on the other foot; to many today, the living truths are found only in the book of science, and the book of Scripture is regarded

---

1. Citations are omitted for quotations from Galileo's *Letter to the Grand Dutchess Christina of Tuscany*, which is reprinted in full in the appendix of this book. When such quotations appear as extracts, they are italicized.

as mythological and irrelevant. Our personal horizons since the time of Galileo have completely changed. Authority has moved from the church (which so dominated everyday life in Galileo's time) to the individual. Many now choose to follow the book of science exclusively, with God beyond the fringe of their horizon. Does science not explain everything? No, there are two realms of knowledge. Everything is not science. Above all, spiritual revelation is not science. As Pope John Paul II elucidates,

> There exist two realms of knowledge, one which has its source in revelation and one which reason can discover by its own power. To the latter belong especially the experimental sciences and philosophy. The distinction between the two realms of knowledge ought not to be understood as opposition.[2]

We refer to these two realms of truth as the two books. For us, as astronomers and Christians, the book of Scripture is the revelation of God to humanity over thousands of years. Whether one accepts these revelations is up to the individual; it depends ultimately on faith, not on bare reason, experiment, or observation (although the faith we are describing does not jettison these either). In contrast, the book of nature encompasses our transient knowledge of science, both observational and theoretical, and its goalposts are ever moving.

Galileo seems to have had a better sense of the two books than his antagonists. He was not threatened by new findings in the book of nature (which may at first appear to contradict the Bible), because Galileo did not see the Bible as a scientific textbook. He saw how progress in the book of nature enables further progress. This is not the role of the book of Scripture: that book emphasizes our place in the universe as spiritual beings and the focus of God's plan for us.

Galileo himself saw the two books as if in balance. He saw the nearby universe with his telescope, and he understood that the Scriptures are about God's relationship with man. In our time, the balance is skewed: the book of nature carries the weight, and the book of Scripture is seen as peripheral or even totally irrelevant.

---

2. John Paul II, "Allocution of the Holy Father John Paul II to the Participants in the Plenary Session of the Pontifical Academy of Sciences," October 31, 1992, in *Papal Addresses to the Pontifical Academy of Sciences 1917–2002*, Pontificiae Academiae Scripta Varia, no. 100 (Vatican City: Pontificia Academia Scientiarum, 2003), 342.

In the book of nature, astronomers find themselves living in a universe that is calculated to be about 92 billion (i.e., thousand million) light years across, filled with billions of stars and galaxies, in which mankind seems insignificant to many (see fig. 1). In contrast, in the book of Scripture, we see mankind sustained by God's grace (his love and undeserved favor toward us). God exists outside space and time; his love is timeless. On the one hand, the book of Scripture does not address all that we can know about space, but on the other hand, it is completely beyond the domain of science to infer that mankind has no central focus in the universe.

God's focus is on his people. The incarnation, God becoming man, is a wondrous sign of spiritual man's focal place in our vast universe. The book of nature is ever changing. As Nigel Brush explains, "From the inside, science does not provide a great deal of confidence in the accuracy and completeness of scientific truth at any one point in time. Far from providing a finished product—the truth and nothing but the truth—science is a work in progress."[3]

In contrast, the world of God's Spirit is not subject to any equations. The book of Scripture is a book with its own context. How can science prove or disprove the revealed grace and love of God? Our receptivity to God and to his Word is inextricably linked to the condition of our hearts as described by Jesus in Matthew 13:3–11 (ESV):

> "A sower went out to sow. And as he sowed, some seeds fell along the path, and the birds came and devoured them. Other seeds fell on rocky ground, where they did not have much soil, and immediately they sprang up, since they had no depth of soil, but when the sun rose they were scorched. And since they had no root, they withered away. Other seeds fell among thorns, and the thorns grew up and choked them. Other seeds fell on good soil, and produced grain, some a hundredfold, some sixty, some thirty. He who has ears [to hear], let him hear."
>
> Then the disciples came and said to him, "Why do you speak to them in parables?" And he answered them, "To you it has been

---

[3] Nigel Brush, *The Limitations of Scientific Truth: Why Science Can't Answer Life's Ultimate Questions* (Grand Rapids, MI: Kregel, 2005), 8.

given to know the secrets of the kingdom of heaven, but to them it has not been given."

Theologians in the early days of modern science were faced with a double dilemma that to some extent is still with us today. On the one hand, there is a dangerous temptation to directly invoke the hand of God when our knowledge in science is limited (the "God of the gaps"). On the other hand, most of these unsolved problems will be solved in the fullness of time, and proposing a divine solution may not in the long run be to the glory of God.

The aurora borealis, or northern lights, is a simple example (see fig. 2): in medieval Europe, before the aurora was scientifically understood, it was thought that heavenly warriors were at work; as a sort of posthumous reward, the soldiers who gave their lives for their king and their country were allowed to battle in the skies forever. There was a gap in our scientific knowledge at the time, and mythologies in the heavens were invoked. Then came a correct scientific understanding of the cause of the aurora borealis involving the sun and the magnetic field of the earth, and the necessity of those heavenly warriors disappeared.

Science is an evolving discipline. Science is never *the* truth but only a set of partial truths. This is the very nature of the scientific method. New observations and new theories develop with time. As Saint Paul writes, "We see through a glass, darkly" (1 Cor. 13:12 KJV).

The Sociology of Science

In Galileo's situation, the discoveries of science *apparently* came into conflict with the literal interpretation of Scripture. What were the theologians to do?

Galileo articulated the problem in his letter:

Those who were expert in astronomy and the natural sciences were convinced by my first announcement, and the doubts of others were gradually allayed unless their scepticism was fed by something other than the unexpected novelty of my discoveries or the fact that they had not had an opportunity to confirm them by their own observations.

Galileo then suggested that his critics would have benefited from listening to an ancient church father:

> *They might have avoided this error [of prescribing the geography of the heavens] if they had paid attention to a salutary warning by St Augustine, on the need for caution in coming to firm conclusions about obscure matters which cannot be readily understood by the use of reason alone.*

Saint Augustine (AD 354–430) suggested that the biblical text should not be interpreted literally if it contradicts what we know from science and our God-given reason. From an important passage in his *De Genesi ad litteram libri duodecim*, or *The Literal Meaning of Genesis* (early fifth century AD), we read,

> It not infrequently happens that something about the earth, about the sky, about other elements of this world, about the motion and rotation or even the magnitude and distances of the stars, about definite eclipses of the sun and moon, about the passage of years and seasons, about the nature of animals, of fruits, of stones, and of other such things, may be known with the greatest certainty by reasoning or by experience, even by one who is not a Christian. It is too disgraceful and ruinous, though, and greatly to be avoided, that he [the non-Christian] should hear a Christian speaking so idiotically on these matters, and as if in accord with Christian writings, that he might say that he could scarcely keep from laughing when he saw how totally in error they are. In view of this and in keeping it in mind constantly while dealing with the book of Genesis, I have, insofar as I was able, explained in detail and set forth for consideration the meanings of obscure passages, taking care not to affirm rashly some one meaning to the prejudice of another and perhaps better explanation.[4]

Augustine's words resonate with us, as they did with Galileo. If the church had heeded Augustine's advice not to impose itself in matters in which it was unskilled, and if power and control had not been such

---

4. Augustine, *The Literal Meaning of Genesis*, trans. and ed. John Hammond Taylor, Ancient Christian Writers 41 (New York: Newman, 1982), 42–43.

a focus for the church at the time, then this long battle between the church and science may never have taken place.

Theologians failed miserably at the time of Galileo in that they misinterpreted Genesis, sprinkling their writings with what Galileo called "vain arguments." Their opinion was that the earth was the center of the universe, and no evidence would change their mind. Gaps in knowledge of matters astronomical were attributed directly to the intervention of God or explained by appealing only to the theologians' own interpretation of verses in the Bible. But as Galileo emphasized, these people show "a greater fondness for their own opinions than for truth." Not even evidence through a telescope would change their minds. In a letter to Johannes Kepler dated 1610, Galileo referred to such people as stubborn "asps":

> My dear Kepler, I wish that we might laugh at the remarkable stupidity of the common herd. What do you have to say about the principal philosophers of this academy who are filled with the stubbornness of an asp and do not want to look at either the planets, the moon or the telescope, even though I have freely and deliberately offered them the opportunity a thousand times? Truly, just as the asp stops its ears, so do these philosophers shut their eyes to the light of truth.[5]

What is an asp? We found Galileo's use of the term "asp" puzzling in the context of stubbornness. After much digging, it became clear that he was talking about a "deaf adder" (Lat. *aspis*).[6]

Serious prejudices against the book of nature often stem from those whose exposure to the scientific method is limited. To be "well grounded in astronomical and physical science" requires as much training as does psychiatry or neuroscience in the medical world. Astrono-

---

5. Galileo, Letter to Johannes Kepler, in *Le Opere di Galileo Galilei: Edizione Nazionale*, ed. Antonio Favaro (1890–1909; repr., Florence: Giunti Barbèra, 2013–2015), 10:423. The original Latin in this letter by Galileo reads, *Volo, mi Keplere, ut rideamus insignem vulgi stultitiam. Quid dices de primariis huius Gimnasii philosophis, qui,* aspidis *pertinacia repleti, nunquam, licet me ultro dedita opera millies offerente, nec Planetas, nec* [crescent moon drawing], *nec perspicillum, videre voluerunt? Verum ut ille aures, sic isti oculos, contra veritatis lucem obturarunt* (emphasis added).

6. According to Isidore, bishop of Seville (AD ca. 560–636), in order to protect itself from snake charmers, an asp would lie down, pressing one ear to the ground, and would curl up its tail to stop (or deafen) sound entering the other ear. The psalmist David vividly describes the scene of an asp stopping its ear in Ps. 58:4–5.

mers would be foolish to pronounce on discoveries in neuroscience or psychiatry; we have not been trained in those specialties. Galileo's letter demonstrates how crucial it is to be thoroughly grounded in astronomy before pronouncing on scientific discoveries. Paraphrasing Augustine's message rather bluntly, don't pontificate about matters that you do not understand.

Galileo's scientific discoveries were never a threat to the book of Scripture, although they were certainly perceived as being so. The book of nature can never be in conflict with the book of Scripture because both have the same author. The one book deals with the universe, the other with God and how he relates to fallen mankind, in need of grace and forgiveness. In the words of Saint Augustine, referring to the book of Scripture, "It was not the intention of the Spirit of God . . . to teach men anything that would not be of use to them for their salvation."[7]

In Galileo's situation, the issue ultimately revolved around authority. The prime agenda of those opposing him was to uphold the geocentric model of the universe as a key to maintaining the power of the church. That was their intention and design. They thus forgot the careful distinction between the realms of the two books as Augustine articulated it.

These tensions reach back to Copernicus, who, in the annals of astronomy, is remembered as the first in Renaissance times to propose a systematic cosmology in which the earth revolves around the sun. His famous treatise *De revolutionibus orbium coelestium* (*On the Revolutions of the Heavenly Spheres*), published in 1543, is often regarded as the birth of modern astronomy and the beginning of the Scientific Revolution. The telescope had not yet been invented: the instruments that Copernicus used were ancient devices like astrolabes, going back to the time of Hipparchus and Ptolemy. Copernicus studied liberal arts in Bologna, then medicine in Padua. In 1497 Copernicus was elected a canon at the cathedral of Frombork in Warmia, and in 1503 he obtained a doctorate in canon law from the University of Ferrara.

---

7. Augustine, quoted in *The Faith of the Early Fathers: A Source-Book of Theological and Historical Passages*, trans. W. A. Jurgens (Collegeville, MN: Liturgical Press, 1979), 3:83.

Although he was not in fact an ordained priest, he was held in high esteem by the Catholic Church, as Galileo emphasized in his letter:

> *They pretend not to know that its author—or rather the one who revived [a moving earth] and confirmed it—was Nicolaus Copernicus, a man who was not just a Catholic but a priest[8] and a canon, and so highly esteemed that he was called to Rome from the furthest reaches of Germany to advise the Lateran Council under Pope Leo X on the revision of the ecclesiastical calendar.*

The church had no problem with celestial measurements and observations, and even with using calculations based on Copernicus's heliocentric model, as long as it could go through the fiction of regarding them as based on a theory, so that they didn't have to face the issues raised by the apparent clash with the text of the Scriptures. This was fine, until Galileo began to promote the Copernican model as fact and forced the church's hand.

Although the scientific establishment of Galileo's day pretended to be objective and only interested in discovering the truth, Galileo perceived the presence of hidden agendas and craven ambitions. Speaking of Copernicus's *On the Revolutions of the Heavenly Spaces*, Galileo wrote,

> *But now that the soundness of its conclusions [a sun-centered world, as proposed by Copernicus] is being confirmed by manifest experiments and necessary demonstrations, there are those who, without even having seen the book, want to reward its author for all his labours by having him declared a heretic—and this solely to satisfy the personal grudge they have conceived for no reason against someone whose only connection with Copernicus is to have endorsed his teachings.*

Galileo's observations ring true today. Just like any other field, science suffers from personal agendas and motivations that can cloud objective reason. Even today it is the personal agenda of some to discredit

---

8. The mistake of calling Copernicus a priest appears to have first been made by Galileo. Copernicus was, in fact, a doctor of canon law but was not ordained. See Edward Rosen, "Copernicus Was Not a Priest," *Proceedings of the American Philosophical Society* 104, no. 6 (1960): 635–61.

the work of others. Some of those held in the highest esteem by the establishment (whether in the sciences or in theology) may try to impede the work of others. Astronomers seeking to publish their research articles may encounter peers appointed by the editor of a journal who may reject an article, only to find those ideas to subsequently emerge as the peer reviewer's own. Priority in scientific discoveries makes for a riveting read. The progress of science can still be modulated by personal agendas, as it was in Galileo's day. The establishment has profound authority, as Galileo points out.

Do the Trees Move the Wind?

Human beings often bring with them great subjectivity, whether they are reading the book of nature or the book of Scripture. In Galileo's day, people hid "under the mantle of false religion and by invoking the authority of Holy Scripture." Their eyes were closed to the book of nature. The Galilean moons orbiting Jupiter implied that there were bodies moving around Jupiter and not the earth; the earth was not the center of the universe. In his letter, Galileo complained about men who use the book of Scripture to discount scientific discoveries. He spoke of those "persisting in their determination to use all imaginable means to destroy me and my works." Judgment was passed on matters scientific by scientifically ignorant theologians who delighted in exposing "heresy." Their goal was to destroy Galileo and everything that was his (including his observations), rather than to contemplate the new astronomy. The book of Scripture was their weapon. Galileo could not accept this. Science in Galileo's mind was never intended to elevate one book above the other. To Galileo, both books had their well-defined foci—as in a coin with two sides, both revealing truth.

Some scientists respond in a similar tone to other scientists who assert their faith in a living God. They fabricate a shield for their atheism out of the mantle and authority of science. The book of nature is their weapon. They use the book of nature to discount the book of Scripture: their eyes are completely shut to grace in the book of Scripture. With Galileo, we find this imbalance wrong. The book of nature was never intended to make judgments on the book of Scripture. The latter is not a scientific textbook. It is a revelation of God's relationship to and redemptive plan for spiritual mankind. At the heart of the incarnation is grace.

Scot Bontrager eloquently pens his thoughts on grace:

> Our natural abilities to discern truth about the world ceases with things invisible—lacking senses to perceive the invisible world there is no way for us to know truths that lead to our eternal beatitude—the perfection for which we were created. The most we can do through natural reason is determine what God is not, but only through discerning God's effects in the world. To get beyond our natural limitations and progress towards our perfection (sharing in the divine nature), we need God's help: an infusion of grace. This grace, for Aquinas, comes in the form of the Holy Scriptures, which are God's willing self-revelation to us. Grace, specifically the grace of revelation as found in the Holy Scriptures, enables a "radical transcendence of the self." Grace, then, is the method and means by which we can come to *know* things necessary for our perfection that we could not know by our natural reason.[9]

In his book titled *Tremendous Trifles*, the literary giant G. K. Chesterton wrote a brilliant essay named "The Wind and the Trees." Chesterton sets the stage:

> I am sitting under tall trees, with a great wind boiling like surf about the tops of them, so that their living load of leaves rocks and roars in something that is at once exultation and agony. . . . The wind tugs at the trees as if it might pluck them root and all out of the earth like tufts of grass. Or, to try yet another desperate figure of speech for this unspeakable energy, the trees are straining and tearing and lashing as if they were a tribe of dragons each tied by the tail. . . . I remember a little boy of my acquaintance who was once walking in Battersea Park under just such torn skies and tossing trees. He did not like the wind at all; it blew in his face too much. . . . After complaining repeatedly of the atmospheric unrest,

---

9. Scot C. Bontrager, "Nature and Grace in the First Question of the *Summa*," *Scot Bontrager* (blog), February 1, 2010, https://www.indievisible.org/Papers/Aquinas%20-%20Nature%20and%20Grace.pdf; italics added.

he said at last to his mother, "Well, why don't you take away the trees, and then it wouldn't wind?"

The great human dogma, then, is that the wind moves the trees. The great human heresy is that the trees move the wind.[10]

There is the invisible world of God's Spirit—the wind—and then there is the material world—the trees—the universe, with its stars and galaxies. Some take the view that the trees move the wind (see fig. 3); they allow science and reason to shape their perspective of God. We would argue otherwise, that the wind moves the trees.

---

10. G. K. Chesterton, "The Wind and the Trees," in *Tremendous Trifles* (1909; repr., London: Methuen, 1930), 61–65.

2

Misunderstanding Truth

A Revulsion to Purpose

In his *Letter to the Grand Duchess Christina*, Galileo wrote,

> Under the cloak of pretended religious zeal, they cite the Holy Scriptures to make them serve their hypocritical purposes, claiming to extend, not to say abuse, the authority of the Scriptures in a way which, if I am not mistaken, is contrary to the intention of the biblical writers and of the Fathers of the Church. They would have us, even in purely scientific questions which are not articles of faith, completely abandon the evidence of our senses and of demonstrative arguments because of a verse of Scripture whose real purpose may well be different from the apparent meaning of the words.

At the time of Galileo's letter, the book of Scripture was the source of all truth, including scientific truths, and Galileo rightly complained that this was out of balance, because it failed to give credence to scientific observation or to recognize how that could be harmonized with Scripture, which always speaks truthfully. In our view, the situation today is equally out of balance, and many scientists today would have us abandon the Scriptures entirely as a source of truth.

Science is driven as much by the *mood of our age* and the personalities and beliefs of individual scientists as it is by empirical data and rigorous theory. Science develops by means of establishing new models, and these models involve physical and mathematical assumptions, which in turn depend on the mood of the age and the style and beliefs of the scientist.

In the words of C. S. Lewis, "We are all, very properly, familiar with the idea that in every age the human mind is deeply influenced by the prevailing temper of mind.... Each [model] reflects the present psychology [mood] of an age almost as much as it reflects the state of that age's knowledge."[1]

Cosmologists of our time have extrapolated the Copernican view, that the earth is not the center of our solar system, to a vastly broader scenario (spanning 92 billion light years)—that our location within the universe of a hundred billion galaxies is in no way special. Our sun is one of a hundred billion stars, located in the outer parts of an apparently random spiral galaxy. Does our location really matter? Does it mean that mankind is a mere accident in an accident of accidents?

But there is something *special* about our universe. A delicate interplay of its age and its size and its fundamental laws has made carbon-based life possible. The law of gravity, and the weak and strong nuclear forces, are finely tuned for the emergence of life. If the force of gravity were stronger, nuclear reactions in the cores of stars would have proceeded so rapidly that their lifetimes would have been very short—too short for the appearance of carbon-based life. Conversely, if the force of gravity were weaker, stars would not have become hot enough for nuclear reactions to start, and we would have no suns. As we understand it, the universe has to be as big as it is—enough time had to pass for the expansion of our universe to have cooled itself off sufficiently after the hot big bang in order for galaxies and stars to form. From our astronomical standpoint, therefore, the expanding universe appears to be relatively old and large.[2]

No reader should be *surprised* that the universe is *so large*, because as far as we can observe, we could not exist in one that is any smaller. The point to be emphasized here is that the fine-tuning makes us scientifically rather special: without this fine-tuning, we would not be here.[3]

---

1. C. S. Lewis, *The Discarded Image: An Introduction to Medieval and Renaissance Literature* (Cambridge: Cambridge University Press, 1964), 222.
2. Charles H. Lineweaver and Tamara M. Davis, "Misconceptions about the Big Bang," *Scientific American* 292, no. 3 (2005): 36–45.
3. For a detailed summary, see David L. Block, *Our Universe: Accident or Design?* (Edinburgh: Scottish Academic Press, 1992).

We should not be dismayed about living in a vast universe, as Bernard le Bovier de Fontenelle was when he wrote, "Behold a universe so immense that I am lost in it. I no longer know where I am. I am just nothing at all. Our world is terrifying in its insignificance."[4]

This is not the mood in the Gospels; they are full of astonishment, wonder, and awe. The incarnation resounds with a central message of purpose. Mankind is special enough that the Creator of this universe visited this world in person out of his love for fallen mankind and died for us. In affirmation, the pre-Copernican medieval models evoked a very positive mood in the theologians of Galileo's day: everything revolved around the earth—mankind was the focal point of creation. On the other hand, Galileo's opponents failed to understand that the incarnation is independent of geographical locale, and positing the earth as moving in its orbit does not denigrate mankind in any way. We wonder why the notion of location was so important to thinkers of that time.

What has changed since the time Galileo wrote his *Letter to the Grand Duchess*? We believe that it is a swing in emphasis and mood. Even over the past 150 years, the discoveries of the new science have led to a change in the empirical experiences that the secular world regards as acceptable to our perception of truth. Some people cannot believe anything that they cannot touch, see, eat, smell, or demonstrate mathematically—Blaise Pascal's *esprit de géométrie* (the geometrical mind), in contrast to his *esprit de finesse* (the intuitive mind). This approach has its place in science, but the level of demonstrable and repeatable perception that is demanded and that has held sway since the nineteenth century has led to blindness to the spiritual world and our current imbalance between the role of the two books.

For Copernicus, the driver of the scientific method was based on "astronomical and geometrical proofs." The book of nature contains multifold mysteries, which requires the eye of experiment and of precise observations to probe it. According to Galileo, Copernicus never embraced the book of Scripture to drive the scientific method; Copernicus was a free-thinking and religious scientist who could

---

4. Bernard le Bovier de Fontenelle, *Conversations with a Lady on the Plurality of Worlds*, or *Entretiens sur la Pluralité des Mondes* (1686), as quoted in Block, *Our Universe: Accident or Design?*, 1.

deeply reflect on the book of Scripture for his spiritual edification and guidance.

Today, however, the driver of the scientific method is not always based, as in Copernicus's day, on precise physical observation. An example here is the hypothesis that instead of one universe, there are myriad others. Why would anyone make such an apparently untestable hypothesis? Astronomers and physicists have recognized that we live in a very finely tuned universe, and questions of a Creator immediately come to the fore. One way out of the finely tuned universe is to invoke a very large number of individual universes, as we explore later. If there are myriad individual universes, then some would argue that there is a finite chance of finding one that is so special. Despite its scientific trappings, belief in a multitude of universes is an exercise in faith, not in observation, but it is then only a small step from there to call God a delusion. We who believe in God and have experienced the love and grace of Jesus would argue otherwise.

As the great physicist Erwin Schrödinger comments,

> Let me briefly mention the notorious atheism of science. . . . Science has to suffer this reproach again and again, but unjustly so. No personal god can form part of a world model that has only become accessible *at the cost of removing everything personal from it*. We know, when God is experienced, this is an event as real as an immediate sense perception or as one's own personality.[5]

We read in the book of Scripture (see fig. 4) that God is indeed spirit (John 4:24). Jesus likened God's Spirit to a wind (John 3:8). We cannot see the wind itself, but we see its effect, bending and swaying the trees, to cite G. K. Chesterton's analogy.

Schrödinger's complaint about a world model that has become accessible only at the cost of removing everything personal from it is a penetrating recognition of the style into which science has plunged itself, and this is a choice that science has made over the past few hundred years. The emphasis on the impersonal and reproducible has its strengths, but a consequence is blindness to the spiritual world.

---

5. Erwin Schrödinger, *Mind and Matter* (Cambridge: Cambridge University Press, 1958), 68; italics added.

Scientists make assumptions in formulating their "objective" pictures of the world and of the universe about us. But as Schrödinger notes, "Every man's world picture is and always remains a construct of his mind," and it is from the minds of people—and not from science—that we are told by some that there is no God.[6] Blindness to the spiritual world is akin to affirming that we hear only the music of Vivaldi without experiencing the spirit of Vivaldi himself.

We affirm with Owen Gingerich that there is the "truth of nature," as described in the book of nature. But to invoke a sense of purpose in the book of nature is, today, a revulsion to many; such is the mood of our age. An "exclusion principle" seems to be at work (science and no God). The book of nature is a book of process, while the book of Scripture is a book of purpose. Gingerich puts it succinctly: not only is there the "truth of nature" but also the "nature of truth" as expounded by Jesus.[7]

"Truth In Her Dress Finds Facts Too Tight"

At one point in his letter, Galileo directly quotes Copernicus, decrying the "idle prattlers" who twist both the words of Scripture and the words of Copernicus:

> *If there should happen to be any idle prattlers who, even though they are entirely ignorant of mathematics, nonetheless take it on themselves to pass judgement in these matters, and dare to criticize and attack this theory of mine because of some passage of Scripture that they have wrongly twisted to their purpose, it is of no consequence to me. . . . Mathematics is written for mathematicians, and if I am not deceived, they will recognize that these labours of mine make a useful contribution to the ecclesiastical state of which Your Holiness now holds the highest office.*

Why was Copernicus on safe ground but not Galileo? Galileo was bold enough to promote Copernicus's theory as fact, which turned out to be a courageous move.

---

6. Schrödinger, *Mind and Matter*, 44.
7. Owen Gingerich, "The Galileo Affair," *Scientific American* 247, no. 2 (1982): 119.

Long before the heliocentric system was condemned in Rome, some preachers took it on themselves to deride Galileo. In his book *Galileo, Science and the Church*, Jerome Langford writes,

> The die was irretrievably cast on December 20, 1614. On that day a Dominican Friar, Father Tommaso Caccini, from the pulpit of Santa Maria Novella in Florence, preached a sermon which strongly condemned the idea of a moving earth. . . . Caccini, a troublesome, ambitious man, left no doubt as to what he thought of the new astronomy or of mathematics in general, for that matter. Galileo wrote several complaints to friends in Rome letting it be known that he was not at all happy to have been the subject of a Sunday sermon. There is no doubt that he had been done a grave injustice. The fact that Father Luigi Maraffi, a Preacher-General of the Dominicans, wrote a formal apology to Galileo did little to placate him.[8]

Though "idle prattlers" like Caccini were committed to misrepresenting Galileo's thought and intentions, Galileo was insistent that scriptural passages should never be "twisted to their purpose" in order to suppress the observations of the natural universe that might appear contrary to Scripture. But equally, the book of Scripture is eternal and requires no scientific validation. Galileo understood this as well. He had no intention of pitting his own scientific research against Catholic piety:

> *My purpose is only that, if in these reflections . . . there is anything . . . useful to Holy Church in reaching a conclusion on the Copernican system, it may be . . . used in whatever way my superiors may decide. . . . Otherwise let this letter [defending the position of Copernicus] be torn up and burnt, for I have no desire for any gain from it which is not in keeping with Catholic piety.*

Galileo offered a balanced view of the two parallel books, both as sources of divine truth (one focused on the scientific, the other focused on the spiritual), but there seems little doubt about which way his reason was drawing him. He knew that the Copernican system was

---

8. Jerome J. Langford, *Galileo, Science and the Church*, Ann Arbor Paperbacks (Ann Arbor, MI: University of Michigan Press, 1992), 55–56.

correct. His telescopic observations had convinced him, and he was driven entirely by his powers of observation and reason.

But we must realize that the book of Scripture requires faith, revelation, and the working of the Holy Spirit to understand its pages. Reason alone is insufficient. We are distinctly reminded of this by our Lord himself, when he said, "However, when He, the Spirit of truth, has come, He will guide you into all truth" (John 16:13 NKJV). "Truth in her dress finds facts too tight," penned Nobel laureate Rabindranath Tagore.[9] The Spirit of truth can never be constrained to human clothing.

An interesting perspective comes from an interview with the late Professor Jean Mesnard of the Sorbonne in Paris (see fig. 5), one of the greatest experts on the mathematician and philosopher Blaise Pascal.[10] Mesnard argued that Galileo's mistake was that he did not see far enough. We see Galileo as the ultimate empiricist. As a faithful member of the Catholic Church, he tried to defend the Scriptures *with reason* against the perceived attack from science and without emphasis on the role of faith, and in doing so, he opened the way to atheism. As we read in the Bible, "Without faith it is impossible to please [God]" (Heb. 11:6). Mesnard stressed that one cannot resort to physical experiments when dealing with God's interaction with our world. Unlike the moons orbiting Jupiter, God cannot be subjected to empirical investigations through a telescope. Reason has its place in interpreting physical experiments, but since the beginnings of the Scientific Revolution in Galileo's time, people have applied reason to evaluating certain spiritual revelations for which reason is not appropriate.

Our conversations with Mesnard drifted down many winding paths, including René Descartes. Unlike Descartes, Pascal addressed the world of the heart: "The heart has its reasons, which reason does not know," penned Pascal.[11]

---

9. Rabindranath Tagore, *Stray Birds* (New York: Macmillan, 1916), 42.

10. In 1654 Pascal experienced a famous and dramatic encounter with God, recorded in his "night of fire," which radically transformed his life and lay quite outside the realm of his reason. Pascal emphasized that God can be found only by the ways taught in the Gospels and not by those of philosophers and scholars. We discuss this further in chapter 12, "Grace in the Life of Blaise Pascal."

11. From Blaise Pascal, *Pensées* (1670), in Blaise Pascal, *Thoughts*, trans. W. F. Trotter, in vol. 48 of *The Harvard Classics*, ed. Charles W. Eliot (New York: Collier, 1910), 99. The original French quote reads thus: "Le cœur a ses raisons, que la raison ne connaît point." Blaise Pascal,

Pascal was read by the cosmologist and astronomer Georges Lemaître in Belgium, who is now officially regarded as the codiscoverer of evidence for the expanding universe. (The vote honoring Lemaître in this way, spearheaded by the International Astronomical Union, took place in October 2018.) In one of his notebooks, Lemaître selected some of the most revealing thoughts penned by Pascal, including this one: "To write against those who made too profound a study of science: Descartes."[12]

Pascal certainly disagreed with Descartes; his entire *Pensées* emphasized revelation and showed the incarnate Word intimately involved with our lives on a daily basis. In another of Pascal's thoughts selected by Lemaître in his handwritten notebook, Pascal stated, "I cannot forgive Descartes. In all his philosophy he would have been quite willing to dispense with God. But he had to make Him give a fillip to set the world in motion; beyond this, he has no further need of God."[13]

At one of our discussions in Paris, Mesnard reflected on Descartes and Pascal as follows: "Descartes flatters himself for proving the existence of God, which seems impossible to Pascal, for whom, however, God is much more alive than for Descartes."

At the heart of the Gospels is the incarnation, which is beyond human reason. But being beyond reason does not mean that the incarnation is false, nor does it mean that reason is irrelevant. Galileo tells us that to truly *understand* is the key. Truth can *never* contradict itself. Reason and revelation are not the same thing, but they are never truly opposed. The book of Scripture is never in contradiction with the book of nature *provided* the sense (or understanding) in both books is correct—aligned to truth and meaning as revealed in both books.

As Galileo wrote,

The reason, then, which they [his critics] give for condemning the view that the Earth moves and the Sun is stationary, is that there are many places in Holy Writ where we read that the Sun moves and the Earth does not; and since Scripture can never lie or be in

---

Pensées, Fragments et Lettres de Blaise Pascal, ed. Armand-Prosper Faugère (Paris: Andrieux, 1844), 2:172.
   12. Blaise Pascal, *Pensées*, trans. A. J. Krailsheimer (New York: Penguin, 1966), 220.
   13. Pascal, *Pensées*, trans. Krailsheimer, 355.

> *error, it necessarily follows that anyone who asserts that the Sun is motionless and the Earth moves must be in error, and such a view must be condemned.*
>
> *The first thing to be said on this point is that it is entirely pious to state, and prudent to affirm, that Holy Scripture can never lie, provided its true meaning has been grasped.*

As the father of modern science, Galileo encouraged theologians to *understand* the meaning of Scriptures. For example, in Job 38:4 we read, "Where were you when I laid the *foundation* of the earth?" (ESV). Does this really mean that the earth rests on concrete pillars or foundations? No, of course not: we know that this would be at variance with observations.

As another example, we read in Job 26:7, "He stretches out the north over the void and hangs the earth on nothing" (ESV). A literal interpretation by one of Galileo's theological contemporaries would be that the earth is hung, like a picture inside a frame, with a piece of string, motionless in space, "on nothing." In contrast, Galileo's telescope revealed a dynamic universe, with four moons orbiting Jupiter, myriad stars spawning the vaults of the Milky Way (see fig. 7), ever-changing phases of the planet Venus, and raging magnetic storms (sunspots) on the surface of our closest star, the sun.

We must interpret what we read in the light of what we know. And what we do know is that the earth is orbiting the sun, and the sun is orbiting the center of our galaxy, and the galaxy itself is rushing through space within the Local Group of galaxies, which in turn is falling in toward the center of the Virgo cluster . . . and the hierarchy continues.

Both books are dynamic: the book of nature is encased in the language of mathematics, and the focus of the Bible is a dynamic relationship between God and his people. The Bible is not a scientific textbook, even as it will ultimately harmonize with science when we understand both books rightly.

3

Understanding the Universe and Scripture

Can Ordinary People Know Great Truth?

In his letter, Galileo expressed the view that the Scriptures were dumbed down to the capacities of the common people, who are "untrained and ignorant," and some parts of the Bible therefore needed exposition by wise expositors to interpret their true meaning.

On the one hand, the Catholic Church at this time had vast powers in deciding who could or could not read and interpret the Scriptures. On the other hand, the Protestant Reformers of the time argued that the book of Scripture did not require the services of wise expositors to penetrate its central meaning. We recall, similarly, that Jesus preached to the masses and called very ordinary common people such as Peter the fisherman and Matthew the tax collector to be his disciples.

The great Bible translator William Tyndale (ca. 1494–1536) well understood that "common people" in his country of birth (England) were ignorant of or "unlearned" in the Scriptures, simply because they did not have any access to the book of Scripture translated into their own language. "Common people" did not read Latin (neither do the majority of our population today). Tyndale's desire was a simple one: for "the boy that driveth the plow to know more of the Scriptures than [the clergy] dost!"[1] This was no small ambition; in Tyndale's time, death sentences for unlicensed possession of Scripture in English were by no means rare.

David Teems gives insight into the veils of obscurity caused by keeping the Bible only in Latin; it was a problem not just for the common people:

> Parish priests in the Middle Ages were often uneducated and could not understand a word of Latin, yet these same priests intoned the Mass regularly. . . . John Hooper, a fellow student with Tyndale at Oxford, became a bishop some years later. . . . Hooper conducted a survey of 311 members of his clergy. Nine priests did not known that there were Ten Commandments; thirty-three had no clue where they were in the Bible (most of them suggested the New Testament); ten could not recite the Lord's Prayer; and thirty did not know Jesus had said it in the first place.[2]

Desiderius Erasmus (ca. 1467–1536) was, if not the greatest scholar of his age, certainly the most famous: no study of the Renaissance or the Reformation would be complete without him. Latin was beyond the "common people," and Erasmus was right on target:

> I totally dissent from those who are unwilling that the Sacred Scriptures, translated into the vulgar [mother] tongue, should be read by private individuals. . . . I wish that the husbandman may sing part of [the Scriptures] at his plough, that the weaver may warble them at his shuttle, that the traveler may with their narratives beguile the weariness of the way.[3]

Teems adds,

> Erasmus is calling for the translation of the Bible into the vernacular; into all mother tongues. Christ should not be hidden. He should not be hoarded or disguised, counterfeited or left unheard. Though His mysteries are impenetrable, the way there, should at least be intelligible, and not just to the select, but to all.[4]

---

1. David Teems, *Tyndale: The Man Who Gave God an English Voice* (Nashville: Thomas Nelson, 2012), 39.
2. Teems, *Tyndale*, 8.
3. Desiderius Erasmus, prologue to the Greek New Testament, quoted in Teems, *Tyndale*, 21.
4. Teems, *Tyndale*, 21.

The age itself was deadly. Teems elaborates: "At the heart of medieval Christianity, if indeed it had a heart, was a reliance on fear and manipulation. The capacity to inspire terror in its faithful was the first rule of order and dominion."[5] To the perceived heretics, there was invariably only one response: *fire*.

George Offor wrote a preface to Tyndale's New Testament of 1526. Offor vividly describes the reign of terror in England at the time:

> Awful were the torments inflicted upon those who . . . dared to read this proscribed book [the New Testament in English]. An aged labourer, father Harding, was seen reading by a wood side, while his more fashionable neighbours were gone to hear mass. His house was broken open, and under the flooring boards were discovered English books of holy scripture; the poor old man was hurried to prison, and thence to the stake, where he was brutally treated, and his body burnt to ashes.[6]

Who do we read was responsible for persecuting the aged father Harding, who was found with our beloved Scriptures translated into English, words pertaining to eternal life? The bishop of Lincoln, John Longland, and Rowland, vicar of Great Wycombe, the bishop's chaplain. John Foxe describes the events that occurred in 1532: "So the next morning came the aforesaid Rowland again, about ten o'clock, with a company of bills and staves, to lead this godly father to his burning."[7]

Sir Thomas More (a councillor to King Henry VIII and lord chancellor from October 1529 to May 1532) did not have gracious words toward Tyndale: "Our Savior will say to Tyndale: Thou art accursed Tyndale, the son of the devil; for neither flesh nor blood hath taught thee these heresies, but thine own father the devil that is in hell."[8]

---

5. Teems, *Tyndale*, xi.
6. George Offor, "Memoir of William Tyndale," in *The New Testament of Our Lord and Saviour Jesus Christ—Published in 1526—Being the First Translation from the Greek into English, by That Eminent Scholar and Martyr William Tyndale; With a Memoir of His Life and Writings, by George Offor* (London: Samuel Bagster, 1836), 29.
7. John Foxe, *Book of Martyrs: The Acts and Monuments of the Church* (London: George Virtue, 1844), 2:264.
8. Offor, "Memoir," 36. The original English reads as follows: "Our Sauiour wyll saye to Tyndale: Thou art accursed Tyndall, the son of the deuyll; for neither fleshe nor bloude hath taught the these heresyes, but thyn owne father the deuyll that is in hell."

It is estimated that between three thousand and six thousand copies of the 1526 edition were printed in Worms, Germany; hidden in bales of cloth, these priceless books of Scripture, in the English tongue, were shipped down the Rhine and smuggled into southeastern England. Almost all copies of Tyndale's New Testament were burned. Men of the cloth involved in the burning of Tyndale's masterpiece in England included Cardinal Thomas Wolsey and the bishop of London, Cuthbert Tunstall (also spelled Tonstall or Tunstal).

Tyndale had many deaf asps or stubborn adders, including Thomas More, who commented that searching for errors in the Tyndale Bible was like searching for water in the sea and who charged Tyndale's translation with having about a thousand translation errors. But we must stress that a significant part of the contention about biblical translation came from fear of the erosion of church authority, social power, and income if the "common people" in England (and elsewhere) could read the Bible in their own language.

To Tyndale, Teems explains, the foundation of the church "was not Simon Peter, but the very faith Peter expressed in his confession [see Matt. 16:15–19]. Upon *this rock, this foundation*, upon *this confession, this revelation*, upon *this thing being revealed to you by my Father* in heaven I will build my congregation, my church."[9]

Tyndale and Galileo shared a common crime: heresy. Tyndale was hunted down in Belgium and strangled and burned at the stake on October 6, 1536. Tyndale had dared to translate the New Testament into English.

There is an interesting parallel here between the tensions of Tyndale and More, on the one hand, and Galileo and Pope Urban VIII, on the other: "It was in William Tyndale that Thomas More saw not only the unquenchable, but worse, the inevitable. He saw the old world, his world, in decline, deflating. He saw a world he had no part of rising in ascent."[10]

We should recognize the difficult political position that Pope Urban VIII was in. As Maurice Finocchiaro elucidates,

---

9. Teems, *Tyndale*, 188.
10. Teems, *Tyndale*, 170.

> A more specific element of religious politics concerns the fact that the climax of the [Galileo] trial in 1632–1633 took place during the so-called Thirty Years War (1618–1648) between Catholics and Protestants. At that particular juncture, Pope Urban VIII, who had earlier been an admirer and supporter of Galileo, was in an especially vulnerable position; thus not only could he not continue to protect Galileo, but he had to use Galileo as a scapegoat to reassert, exhibit, and test his authority and power.[11]

Finocchiaro shows brilliant insight here: at the heart of the Galileo trial was *not science* but rather a potential tsunami eroding church power and authority—beginning, first and foremost, in the mind of Pope Urban VIII himself, in those tumultuous religious-political times.

With regard to only a privileged few understanding the book of Scripture, we would distance ourselves from the disdain with which Galileo regarded the common person. He spoke of the masses as "untrained and ignorant," and concluded that "it is necessary for wise expositors to explain [the Scriptures'] true meaning to those few who deserve to be set apart from the common herd, and to point out the particular reasons why they have been expressed in the terms that they have."

Erasmus and Tyndale would have most fervently disagreed with Galileo on this point. Tyndale devoted his life to providing access to the Bible by those who were not learned in Latin. His rendition of the New Testament into English is a masterpiece. Every person reading it could understand God's plan for his or her salvation. It is a tribute to Tyndale that, in 1611, the fifty-four independent scholars who created the King James Version of the Bible drew significantly from Tyndale (as well as subsequent translations that descended from his). It has been estimated that 83 percent of the New Testament and 76 percent of the Old Testament in the King James Version is Tyndale's.

As part of his translations, Tyndale crafted such familiar phrases as these:

> "lead us not into temptation but deliver us from evil"
> "knock and it shall be opened unto you"

---

11. Maurice A. Finocchiaro, trans. and ed., *The Trial of Galileo: Essential Documents* (Indianapolis: Hackett, 2014), 7–8.

"twinkling of an eye"
"a moment in time"
"fashion not yourselves to the world"
"seek and ye shall find"
"ask and it shall be given you"
"judge not that ye be not judged"
"the word of God which liveth and lasteth forever"
"let there be light"
"the powers that be"
"my brother's keeper"
"the salt of the earth"
"a law unto themselves"
"filthy lucre"
"it came to pass"
"gave up the ghost"
"the signs of the times"
"the spirit is willing, but the flesh is weak"
"fight the good fight"

These phrases are readily understood by all, as they were for the aged laborer father Harding. Let us remember that the appearance of the King James Bible in 1611 was only one year after Galileo's *Sidereus Nuncius* had been published in Venice, in 1610. As noted above, the church in Rome was under fire from both theological and scientific directions: a new and less controllable worldview was rising.

Nevertheless, Galileo was incorrect in his assumptions about the Bible and ordinary people. For example, he wrote,

> *From this it seems reasonable to deduce that whenever Scripture has had occasion to speak about matters of natural science [such as astronomy], especially those which are obscure and difficult to understand, it has followed this rule so as not to cause confusion among the common people and make them more sceptical of its teaching about higher mysteries.... The primary intention of Holy Writ [is] divine worship and the salvation of souls, and matters far removed from the understanding of the masses.*

As Galileo said, the Bible has not hesitated to obscure some pronouncements. The book of nature and the book of Scripture both have their mysteries that cannot be lightly shrugged off. However, the use of physical imagery in the Scriptures, of the heavens as a "curtain" or "tent," should not be taken literally: it is rich in metaphor and helps us picture the vastness of the starry vaults or canopy of our night sky. And the common people were able to grasp far more than Galileo gave them credit for.

Galileo pointed out that mysteries are encountered in the physical sciences as well. We read of "God making the stars," and we accept that he created everything, as stated in the Nicene Creed, but precisely how should this theological statement be interpreted in a scientific context without compromising either professed belief? What are the actual physical processes triggering the formation of stars (see fig. 8) on small, intermediate, and large scales? There are a variety of possible scenarios—from the compression of clouds of gas to expanding rings of gas to the triggering of stellar birth via supernovae (or exploding stars) to the roles induced by elongated features in many galaxies (known as bars), just to name a few proposals. Entire conferences continue to focus on the mechanisms that trigger the formation of stars. Bruce Elmegreen writes, "A variety of processes cause interstellar gas to become cold enough and dense enough to form stars. . . . On galactic scales, stellar instabilities, spiral waves, and global perturbations like bars can move the gas around" to trigger the birth of stars.[12]

But that does not imply that the book of nature should be tossed away. Galileo was not surprised that such mysteries should be encountered in the book of nature as well. We must understand how incomplete the book of nature is today and how fast it is changing. Imagine what scientists a hundred years in the future will think of our state of knowledge. How many astronomy books, written in the 1800s, do we use as reference material for students in our classrooms today?

As for salvation in the book of Scripture, a fisherman need not understand or even contemplate the concept of star formation. All

---

12. Bruce Elmegreen, quoted in "Star Formation in Galaxies," in *The Spectral Energy Distribution of Galaxies: Proceedings of the 284th Symposium of the International Astronomical Union, Held at the University of Central Lancashire, Preston, United Kingdom, September 5–9, 2011*, ed. Richard J. Tuffs and Cristina C. Popescu (Cambridge: Cambridge University Press, 2012), 317–29.

that is needed is a personal encounter with Jesus. We would argue that salvation of souls is infinitely *within reach* of the comprehension of the common people, such as Peter the fisherman, who in his common humanity thrice denied Jesus at his trial.

In his letter, Galileo now moved to the crux of the two books: they cannot be mutually exclusive because they have the same author.

Secular Science: The New Cathedral

In line with Tyndale's efforts with the Scriptures, Galileo understood the need for being a great popularizer of science, by writing books that could be easily digested by those who were not astronomers themselves. His language appeals to the modern ear (although written centuries ago). The translations of Galileo's books into modern-day language are still fresh and forceful.

Many scholars today, too, recognize the need to popularize their ideas by writing books aimed at the general public. One of the most influential authors guiding atheistic beliefs among the masses today is biologist Richard Dawkins, at Oxford, author of several best-selling books. One of his bold assertions is, "Our own existence once presented the greatest of all mysteries, but that is a mystery no longer because it is solved."[13] Dawkins furthermore asserts, "The truth is more magical—in the best and most exciting sense of the word—than any myth or made-up mystery or miracle. Science has its own magic: the magic of reality."[14] The implication by Dawkins is that science is reality. God, he asserts, is nonreality, or a delusion.

We believe that the claims by Dawkins are an overly optimistic view of the domain of science. The scientific method has as its focus only "the truth of nature" and not "the nature of truth."

With regard to the power and domains of the scientific method, we offer two important comments. First, by no stretch of the imagination has everything been solved by science. From a cosmological perspective, over 95 percent of the universe is not visible: it is in the form of dark matter and dark energy, about which we know hardly

---

13. Richard Dawkins, *The Blind Watchmaker* (Harlow: Longman Scientific and Technical, 1986), xiii.
14. Richard Dawkins, *The Magic of Reality: How We Know What's Really True* (New York: Simon and Schuster, 2011), 266.

anything more than that they exist. The stars that astronomers see in spiral galaxies, for example, merely constitute a fraction of their total mass; the disks of spiral galaxies are immersed in extensive envelopes of dark matter—matter that neither emits nor absorbs light. Several decades ago, one of us (Freeman) noted that "there must be in these galaxies additional matter which is undetected. . . . Its mass must be at least as large as the mass of the detected galaxy."[15] Without this dark matter, galaxies may not have been able to form, and we would not be here.

Not only is there the enigmatic dark matter problem. Certain exploding stars (known technically as type SN Ia[16]) in distant galaxies are, on average, fainter—and therefore farther from us—than predicted. The universe is accelerating! The term *dark energy* is used to account for the repulsive force fueling this acceleration.

Second, even if we understood such matters (dark matter and dark energy), science cannot solve the central mystery: "Why are we here?" Cosmologist George Ellis affirms this thinking:

> When science studies the nature of cosmology, it does so on the basis of the specific laws of physics that apply in the unique Universe we inhabit. It can interrogate the nature of those laws, but not the reason for their existence, nor why they take the particular form they do. Neither can science examine the reason for the existence of the Universe. These are metaphysical issues, whose examination lies outside the competence of science *per se*.[17]

The need for well-informed minds, free of the fetters of myths, is so crucial in our exploits for truth. Teems sums this up eloquently:

---

15. K. C. Freeman, "On the Disks of Spiral and S0 galaxies," *Astrophysical Journal* 160 (1970): 828.

16. Nobel laureate Professor Brian P. Schmidt, in his 2011 Nobel address, states, "Supernovae (SN), the highly luminous and physically transformational explosions of stars, show great variety, which has lead to a complex taxonomy. They have historically been divided into two types based on their spectra. Type I supernovae show no hydrogen spectroscopic lines, whereas Type II supernovae have hydrogen. Over time, these two classes have been further divided into sub-classes. The Type I class is made up of the silicon rich Type Ia, the helium rich Type Ib, and the objects which have neither silicon nor helium in abundance, Type Ic." "The Path to Measuring an Accelerating Universe," Nobel lecture, December 8, 2011, https://assets.nobelprize.org/uploads/2018/06/schmidt_lecture.pdf?_ga=2.254506317.451573194.1535734006-218083626.1535734006.

17. George F. R. Ellis, "The Thinking Underlying the New 'Scientific' World-Views," in *Evolutionary and Molecular Biology: Scientific Perspectives on Divine Action*, ed. Robert John Russell, William R. Stoeger, and Francisco J. Ayala (Vatican: Vatican Observatory Foundation, 1998), 254.

According to the scriptures, creation was set to order by a single voice, at the expense of one clear sentence, the brief prelude to all existence: *Dixitque Deus fiat lux et facta est lux.* "God said, let there be light, and there was light" (Gen. 1:3). In a very real sense, light is the argument Galileo presses. It is what he represents for us today. Clarity. Lucidity. Light. Reading and interpreting the unknowns that exist, not by the light of fable or superstition, but with the well-informed mind, free of its fetters.[18]

The beginning of the universe has occupied the minds of some of our greatest cosmologists. Has everything now been solved? Is secular science indeed the new cathedral, as believed by multitudes? In a review published several decades ago, one of us (Block) noted, "In such a model the actual creation process at t=0 cannot therefore be observed."[19]

The use of a few key words here is important: the science of cosmology operates by means of mathematical *models* in vogue at the time, and such mathematical models have their *assumptions*. Furthermore, the *actual creation process* at the very beginning cannot be observed through any telescope. It will forever be masked. As in any model, assumptions have to be made. The point, however, is this: it is incorrect to state that science deals with all aspects of our existence or why the universe exists in the first place.

One of the greatest cosmologists of all time was the late Stephen Hawking, Lucasian Professor of Mathematics at the University of Cambridge. He believed he had expunged God. In his mathematical models, Hawking found no need for a Creator, but Lord Martin Rees (a former president of the Royal Society of London) sets such claims in perspective:

> His name [Hawking] will live in the annals of science; millions have had their cosmic horizons widened by his best-selling books; and even more, around the world, have been inspired by a unique example of achievement against all the odds—a manifestation of amazing will-power and determination. . . . He had robust common sense, and was ready to express forceful political opinions.

---

18. David Teems, email to David Block, July 8, 2016.
19. David Lazar Block, "General Relativity, and Its Applications to Selected Astrophysical and Cosmological Topics," *Quarterly Journal of the Royal Astronomical Society* 15 (1974): 266.

However, a downside of his iconic status was that his comments attracted exaggerated attention even on topics where he had no special expertise.[20]

Lord Rees strikes the same note of caution as does George Ellis: it is outside the domain of science to claim that God is irrelevant.

We salute Galileo for not following the mood of his age but, at the same time, for being courageous enough to embrace both God and science. He accepted the difference between impersonal nature, not caring a whit whether we understand her laws, and God's interaction with spiritual man, along with the enormous efforts that God has gone to in revealing himself through the book of Scripture. In the book of Scripture, God gives the clearest revelation of himself, while the book of nature, which also reveals God in general ways, primarily exposes us to the world of experience and experiment.

Where the road of science will take us is unknown; our progress along this road is limited by our confinement to space and time and by our very limited understanding of that which we observe. Dark matter, dark energy, and why spiral galaxies have the structures they do are but three examples. Each era has its scientists who believe that their understanding is fast approaching the end of the road, when the end is, in reality, merely a bend. This was strikingly portrayed by the astronomer and mathematician Simon Newcomb (1835–1909), who over a century ago, in 1888, is believed to have said, "We are probably nearing the limit of all we can know about astronomy."[21]

In 1903, a similar sentiment was expressed by the famous physicist A. A. Michelson. He wrote, "The more important fundamental laws and facts of physical science have all been discovered, and these are so firmly established that the possibility of their ever being supplanted in consequence of new discoveries is exceedingly remote."[22]

---

20. Lord Martin Rees, "Professor Stephen Hawking: An Appreciation," Trinity College Cambridge, March 14, 2018, https://www.trin.cam.ac.uk/news/professor-stephen-hawking-an-appreciation-by-lord-rees/.

21. This very widely quoted comment is attributed to Newcomb. See, for example, Prasenjit Saha and Paul A. Taylor, *The Astronomer's Magic Envelope* (New York: Oxford University Press, 2018), 4.

22. Albert Abraham Michelson, *Light Waves and Their Uses* (Chicago: University of Chicago Press, 1903), 23–24. See also I. Bernard Cohen, *Revolution in Science* (Cambridge, MA: Belknap Press of Harvard University Press, 1985), 281.

This was just before Einstein published the special theory of relativity in 1905, followed by his general theory of relativity a few years later. Einstein had predicted that the light from stars would be deflected when passing around our sun; this was confirmed at an eclipse of the sun in 1919. We read,

> Thus the results of the expeditions to Sobral [in northern Brazil] and Principe [an island off the coast of West Africa] can leave little doubt that a deflection of light takes place in the neighbourhood of the sun and that it is of the amount demanded by Einstein's generalized theory of relativity, as attributable to the sun's gravitational field.[23]

The year 1919 was a mere sixteen years after the statement by Michelson. We recall, too, that quantum theory was coming on stage around the same time: the famous Fifth Solvay Conference, titled "Electrons and Photons," was held in Belgium in 1927 and celebrated the wide acceptance of quantum theory.

We see a universe of purpose: a universe unfolding, in our view, over fourteen billion years of the richest of histories, with the final goal of the universe containing his people, people of purpose. Galileo argued that God is not less excellently revealed in nature's actions (see Ps. 19:1: "The heavens declare the glory of God" [ESV]). Although some scientists believe that the world is devoid of purpose and that all our thoughts are accidents of environment, we would concur with G. K. Chesterton: "The man who represents all thought as an accident of environment is simply smashing and discrediting all his own thoughts—including that one."[24]

Galileo emphasized that the Bible is not chained in its expression to conditions as strict as those governing physical laws. We see God revealed by the truths contained in the Scriptures and displayed in his

---

23. F. W. Dyson, A. S. Eddington, and C. Davidson, "A Determination of the Deflection of Light by the Sun's Gravitational Field, from Observations Made at the Total Eclipse of May 29, 1919," in *Philosophical Transactions of the Royal Society of London*, ser. A, vol. 220 (London: Harrison and Sons, 1920), 332. J. J. Thomson concluded that this confirmation was the most important result in connection with the theory of gravitation since the time of Sir Isaac Newton (*The Times*, London, November 7, 1919). Overnight, Einstein became the most famous scientist in the world.
24. G. K. Chesterton, "The Wind and the Trees," in *Tremendous Trifles* (1909; repr., London: Methuen, 1930), 64.

role as executor, as in the first chapter of John. What does Scripture primarily reveal? Galileo explained:

> *I believe therefore that the purpose of the authority of Holy Scripture is chiefly to persuade men of those articles and propositions which, being beyond the scope of human reasoning, could not be made credible to us by science or by any other means, but only through the mouth of the Holy Spirit.*

What was Galileo thinking of? The incarnation of our Lord is beyond the scope of human reasoning, for example. As Chesterton wrote in *The Everlasting Man*,

> Right in the middle of all these things stands up an enormous exception. It is quite unlike anything else. It is a thing final like the trump of doom, though it is also a piece of good news; or news that seems too good to be true. It is nothing less than the loud assertion that this mysterious maker of the world has visited his world in person. It declares that really and even recently, or right in the middle of historic times, there did walk into the world this original invisible being; about whom the thinkers make theories and the mythologists hand down myths; the Man Who Made the World. That such a higher personality exists behind all things had indeed always been implied by all the best thinkers, as well as by all the most beautiful legends. But nothing of this sort had ever been implied in any of them. It is simply false to say that the other sages and heroes had claimed to be that mysterious master and maker, of whom the world had dreamed and disputed. Not one of them had ever claimed to be anything of the sort. Not one of their sects or schools had even claimed that they had claimed to be anything of the sort. The most that any religious prophet had said was that he was the true servant of such a being. The most that any visionary had ever said was that men might catch glimpses of the glory of that spiritual being; or much more often of lesser spiritual beings. The most that any primitive myth had even suggested was that the Creator was present at the Creation. But that the Creator was present at scenes a little subsequent to the supper-parties of Horace, and talked with tax-collectors and government officials in

the detailed daily life of the Roman Empire, and that this fact continued to be firmly asserted by the whole of that great civilisation for more than a thousand years—that is something utterly unlike anything else in nature. It is the one great startling statement that man has made since he spoke his first articulate word, instead of barking like a dog. Its unique character can be used as an argument against it as well as for it. It would be easy to concentrate on it as a case of isolated insanity; but it makes nothing but dust and nonsense of comparative religion.[25]

We understand this through the lens of the Holy Spirit. But now we have come too far. To many, science is the new cathedral. We are very impressed by our own reason and intellect, with not much room for revelation, the book of Scripture, and God becoming man. The pendulum has swung 180 degrees.

---

25. G. K. Chesterton, *The Everlasting Man* (New York: Image Books, 1955), 265–66.

4

What Grace and Space Cannot Tell Us

Known Unknowns

In 2015, we celebrated the centenary of the publication of Albert Einstein's theory of general relativity. Not only was Einstein a great physicist, but he also had great admiration for the person of Jesus as described in the Gospels. An excerpt of an interview with Einstein by George Viereck offers valuable insight into the nature of truth:

> Viereck: "To what extent are you influenced by Christianity?"
> Einstein: "As a child I received instruction both in the Bible and in the Talmud. I am a Jew, but I am enthralled by the luminous figure of the Nazarene."
> Viereck: "Have you read Emil Ludwig's book on Jesus?"
> Einstein: "Emil Ludwig's Jesus is shallow. Jesus is too colossal for the pen of phrase-mongers, however artful. No man can dispose of Christianity with a *bon mot*."
> Viereck: "You accept the historical existence of Jesus?"
> Einstein: 'Unquestionably! No one can read the Gospels without feeling the actual presence of Jesus. His personality pulsates in every word. No myth is filled with such life."[1]

---

1. "What Life Means to Einstein: An Interview by George Sylvester Viereck," *Saturday Evening Post*, October 26, 1929, 117, http://www.saturdayeveningpost.com/wp-content/uploads/satevepost/einstein.pdf.

Neither Einstein nor Galileo believed that his research pushed him away from divine things. To Einstein, God is not personal but rather an "infinitely superior spirit that reveals itself in the little that we, with our weak and transitory understanding, can comprehend of reality."[2] Galileo believed, with Saint Augustine, that there is ultimately no genuine conflict between the book of grace and the book of science.

If only the church of Galileo's time had heeded the words of Saint Augustine:

> If anyone shall set the authority of Holy Writ against clear and manifest reason, he who does this knows not what he has undertaken; for he opposes to the truth not the meaning of the Bible, which is beyond his comprehension, but rather his own interpretation; not what is in the Bible, but what he has found in himself and imagines to be there.

If the church had listened to Augustine, the trial of Galileo may have taken a completely different turn—in fact, it may never have occurred.

Galileo invoked the merit of "solid reasons and experiences of human knowledge," but we must comment that the sciences nevertheless pursue an ever-winding and rocky road toward truth—never actually getting there. In retrospect, with the benefits of viewing the four centuries of scientific progress after Galileo, we think that Galileo was overconfident about the powers of human reason. As scientists, we are skeptical about "clear and manifest reason."

The crisis in physics around the turn of the nineteenth to the twentieth century shows the limitations of science and human reasoning only too clearly. We are referring here to the events that overturned much of the confident classical physics of the nineteenth century and led us to quantum theory and relativity. Truth, however, never contradicts itself, as Galileo pointed out:

> *For who can place limits on the human mind, or claim that we already know all that there is to be known? Will it be those who*

---

2. See, for example, Walter Isaacson, *Einstein: His Life and Universe* (New York: Simon & Schuster, 2007), 388.

on other occasions admit, quite rightly, that "What we know is only a tiny part of what we do not know"?

Indeed, since we have it from the mouth of the Holy Spirit that "He has given up the world to disputations, so that no man may find out what God made from the beginning to the end," I do not think we should contradict this by closing the path to free speculation concerning the natural world, as if everything had already been discovered and revealed with absolute certainty. Nor do I think it should be considered presumptuous to challenge opinions which were formerly commonplace, or that anyone should be indignant if someone does not share their opinion on a matter of scientific dispute.

At this point, let us ponder the famous wise words of Donald Henry Rumsfeld, who served as US Secretary of Defense from 1975 to 1977 and again from 2001 to 2006. In February 2002 Rumsfeld stated at a Defense Department briefing,

> There are known knowns. These are things we know that we know. There are known unknowns. That is to say, there are things that we know we don't know. But there are also unknown unknowns. There are things we don't know we don't know.[3]

How drastically has our understanding of cosmology and physics changed over the past fifty years. Physicists are hard at work to develop a grand unified theory, which operates on both the macroscopic and microscopic scales. As noted earlier, dark matter and dark energy have entered center stage, as known unknowns: they are recent concepts in our inventory of knowledge—1930s for dark matter and 1990s for dark energy—but will they still dominate the minds of astronomers and physicists in twenty years' time, or will other fundamental insights (as yet unknown) spawn a new scientific revolution by then? Galileo was right on the mark when he stated that "what we know is only a tiny part of what we do not know."

Rumsfeld could have added a fourth term: the "unknown knowns," things that we don't know that we know, like the knowledge in the

---

3. See, for example, David C. Logan, "Known Knowns, Known Unknowns, Unknown Unknowns, and the Propagation of Scientific Enquiry," *Journal of Experimental Botany* 60, no. 3 (2009): 712–14.

back of our minds, acquired and long forgotten, that can guide our behavior in unforeseen situations. As Pascal wrote, there are "four kinds of persons: zeal without knowledge; knowledge without zeal; neither knowledge nor zeal; both zeal and knowledge."[4]

There is, of course, a world of knowledge unknown to us but known to those with whom we cannot communicate. We think of the instinct of kangaroos triggered by the approaching and devastating Canberra bushfires of 2003; the kangaroos knew what to do a full twenty-four hours before the blazing fires actually swept over Canberra. Their knowledge is intuitive. Extreme danger was at hand: their intuition drove them to vacate the area.

Is the kangaroo's intuition a known known or an unknown known? We should ask a kangaroo. The kangaroo's intuitive knowledge is in no way based on intellectual or analytic capacity or "solid reasons." There is another world of knowledge out there. Where do these unknown knowns come from? Where would our kangaroo fit into Pascal's classification?

Unlike Galileo's time, some scientists today may go one step further and promote atheism under the mantle of science. Secular science is to many the new cathedral, as discussed earlier. Atheism and its scientific pretensions would be an example of a known known.

The Search for Truth and Purpose

But returning to the nature of truth and the truth of nature, the book of Scripture does not explicitly demand that the universe is only six thousand years old (in our opinion, science leads us to conclude that it is fourteen billion years old). Equally so, the creation of mankind is highly complex, and the physical recipe cannot be found in the book of Scripture—thus, the appearance of mankind must be understood in terms of the ethos of *purpose and salvation*.

In the making of mankind, evolution is the major scientific theme of our time, and God's role in the appearance of humanity is often seen as unnecessary. We use our cognitive faculties to understand (or

---

4. Blaise Pascal, *Pensées*, ed. and trans. Roger Ariew (Indianapolis: Hackett, 2005), 157. This *Pensées* fragment is numbered 867 in the translation Pascal, *Pensées*, introduction by T. S. Eliot (New York: E. P. Dutton, 1958).

to try to understand) the universe. Some may claim that our brains (and our beliefs) are selected for adaptation and for survival—and not for truth—and therefore our beliefs are transitory. But we can imagine that the belief systems of a person painting on a cave wall long ago, seeking shelter from ravaging animals outside, would not be so different *in the search for purpose and truth* from those of a person sitting today in an apartment in New York.

This was markedly brought to the fore by a recent visit to one of the very remote San villages near the Tsodilo hills in Botswana. For thousands of years, the San were hunter-gatherers; some were artists. The Tsodilo hills are today a world heritage site, containing thousands of rock art paintings. At those hills, we met a member of the San, named |Xontae. In his quest for purpose, |Xontae has experienced the grace of Jesus. The search for purpose is universal.

Galileo spoke of opinions that have become common. In the days of G. K. Chesterton, the philosophy of solipsism (that a person believes in his own existence but not in anybody or anything else) may have been common, but was it true? Yet if our cognitive faculties are founded on truth, the pendulum swings 180 degrees. Truth—in particular, he who is "the truth"—never changes with time (John 1:14–18).

The belief that pleases many other people best—that of a "hands-off" God (or no God)—makes no sense to us; that would be contrary to everything we read in the Gospels, of sin, forgiveness, prayer, miracles, and the grace of Jesus. We do not understand the mechanism as to how God breathed his spirit into mankind (anymore than we know what triggers the formation of stars), but we fully accept that God is intimately involved in our lives, as poignantly penned in the first chapter of John. Just after Galileo's death in 1642, Pascal's "night of fire," in 1654, resounded with the truth of Pascal meeting the hands-on Creator of the universe face-to-face (see chap. 12).

Galileo issues a healthy challenge for scientists in the twenty-first century:

> *No one should be scorned in physical disputes for not holding to the opinions which happen to please other people best, especially*

concerning problems which have been debated among the greatest philosophers for thousands of years.

We know that paradigm shifts are often bitter and difficult for those whose worldviews are being swept away, but Galileo's experiences with his critics are truly remarkable. It is extraordinary to think that the Bible could be invoked to disprove that Jupiter had four moons moving in orbits around it, as revealed by Galileo's eyes as he peered through his telescopes.

There are those who distort the book of Scripture to their own ends; such distortions still happen today. Galileo's paragraph above again stresses the need for care in considering what the two books actually address. We affirm God's passion for the world and his ability to intervene in the book of nature (such as Jesus entering the domains of time and of space in our world), but that doesn't mean that Scripture tells us everything we can learn about science—it doesn't aim to do so. Similarly, we cannot perceive scriptural truth with our spiritual eyes apart from revelation and grace.

The poet William Blake sums it up as follows:

This Lifes dim Windows of the Soul
Distorts the Heavens from Pole to Pole
And leads you to Believe a Lie
When you see with not thro the Eye.[5]

As Galileo knew, the danger of ignorance in these matters is compounded by pretenses to authority. Of his critics, Galileo wrote, quoting Jerome, that they

gratify people's ears with carefully constructed phrases, and think that whatever they say is the word of God, without bothering to find out what the prophets and apostles taught.

Some with theological or political authority and no experience in science are ready to make judgments on the goals, methods, and conclusions of science. Instead, such individuals would be wise to adorn

---

5. William Blake, "The Everlasting Gospel," in *The Complete Poetry and Prose of William Blake*, ed. David V. Erdman (Berkeley: University of California Press, 1981), 520.

themselves with caution and humility in matters outside their realm of expertise.

Galileo explored this further by speaking of people who expounded to others what they themselves did not understand, this time in the Scriptures. In the words of Saint Jerome (347–420),

> The art of interpreting the Scriptures is the only one of which all men everywhere claim to be the masters. . . . They do not deign to notice what the prophets and apostles have intended, but they adapt conflicting passages to suit their own meaning as if it were a grand way of teaching—and not rather the faultiest of all—to misrepresent a writer's views and to force the Scriptures reluctantly to do their will.[6]

Likewise today, some with intellectual authority and little understanding of revelation in theology are equally ready to pronounce that God is a delusion. Crucial in this context is the *depth* to which both books are properly understood; Galileo was eager that we need not concern ourselves with the shallowness of people who claim to have *authority* in the book of Scripture but in reality have only a superficial understanding of it. Galileo warned against those who, "prompted by pride, bandy fine-sounding words as they hold forth about Holy Writ."

However, in science, too, we dare not be led by pride or heavy words; moreover, we dare not have personal philosophical agendas within science and claim those as scientific truth. As noted earlier, it is very easy for atheism to masquerade under the mantle of science.

---

6. Jerome, "Letter LIII: To Paulinus," in *The Letters of St. Jerome* (London: Aeterna, 2016), 162.

5

The Fraud of Scientism

The Modern Bias against Theology

Galileo wrote,

> I fear there may be some cause for confusion if the pre-eminence which entitles theology to be called the queen of sciences is not clearly defined. It could be because the material taught by all the other sciences is encompassed and demonstrated in theology, but by more comprehensive methods and with more profound learning. . . . Or it could be because the subject matter of theology surpasses in dignity the subject matter of the other sciences, and because it proceeds by more sublime methods. I do not think that theologians who are conversant with the other sciences would claim that theology deserves to be called queen for the first of these reasons, for it is hard to believe that any of them would say that geometry, astronomy, music and medicine are more comprehensively and precisely expounded in Scripture than in the works of Archimedes, Ptolemy, Boethius and Galen. It follows that the regal pre-eminence of theology must be of the second kind, namely on account of its elevated subject matter, its marvellous teaching of divine revelation, which human comprehension could not absorb in any other way, and its supreme concern with how we gain eternal beatitude. . . . And if theology is concerned with the most elevated contemplation of the divine, then those who practise and

profess it should not claim the authority to lay down the law in fields where they have neither practised nor studied. If they did, they would be like an absolute prince who, knowing he was free to command obedience as he wished, insisted that medical treatment be carried out and buildings be constructed as he dictated even though he was himself neither a doctor nor an architect, thereby causing grave danger to the lives of his unfortunate patients and the evident ruin of his buildings.

Galileo expressed his frustrations with the hierarchy of disciplines. In his time, theological ideas were supreme, and theology was indeed the queen of all the sciences. Modern science was just beginning to emerge as a subordinate discipline, strongly constrained by the entrenched power of Galileo's theological peers. Galileo gives us a bitter insight into the hierarchical intellectual structure of his day. In our time, the balance has swung, and science itself is the queen of the sciences. Science is advanced by those with scientific expertise. While there may be some background noise from theological perspectives, it does not really affect the progress of modern science. Some of our scientific colleagues, blind to scriptural revelation, are only too ready to trash theology as an inferior and worthless discipline, devoid of all truth. This is the ideology of scientism—an optimistic faith in the power of science to resolve the mysteries of the world, a form of science that dismisses God, who entered our world, as a fairy tale.

Galileo's time was a time of transition. Theology and the Scriptures were still the framework for interpretation of physical phenomena, though that was changing. Theologians at the time of Galileo were trying to force the universe to fit their interpretation of Scripture by quoting and twisting certain passages without recourse to scientific knowledge. To illustrate their closed minds, many were unwilling even to look for themselves through Galileo's spyglasses or telescopes. Galileo railed at these constraints and particularly at the arrogant and scientifically inept commentary and criticism that he had to endure.

If we wish to understand the physics of our complex universe, we undertake rigorous training in those disciplines ourselves, and we collaborate with other scientists. We are fully aware that this is not the role of the book of Scripture. The book of Scripture has a pivotal role in the spiritual side of our being and purpose in this world, and we do not see the book of Scripture threatened by the book of nature. For our spiritual health and reconciliation with God, we do not seek a set of equations. We seek a person, the Logos, the incarnation. The book of Scripture is focused on Jesus, who holds the key to our salvation.

Galileo accepted the regal dignity of theology, on the basis of its subject and "its marvellous teaching of divine revelation, which human comprehension could not absorb in any other way, and its supreme concern with how we gain eternal beatitude." But he was less convinced about its practitioners—he stressed that competence in theology does not bestow authority in technical areas in which one is not expert. Humility is called for, whether it be in theology or in science. God works not only in the heights but also in the lowest depths. It was not in a regal palace but in a cave in Bethlehem that Jesus was born.

Galileo then raised the question of intellectual integrity in the face of the political pressure of the time. He wrote,

> *Then, to command that professors of astronomy should be responsible for undermining their own observations and proofs as no more than fallacies and false arguments, is to command something quite impossible for them to do. For it amounts to telling them not to see what they see, and not to understand what they understand and, indeed, to find in their research the very opposite of what evidence shows them.*

Observers then and now must report what they see, not what the politicians, scientists, and theologians of the day would like them to see. We are on the road to truth. But each epoch of time has its mood, and scientists who may go against the mood of the current age (e.g., by invoking purpose) may be unpopular. The pressures of political correctness now are not so different from the pressures of theological correctness faced by Galileo, although the potential outcomes in Galileo's time were different. It would be unthinkable then and now for scientists to claim under pressure that they do not see something when they do in fact see it. On the other hand, they have a grave responsibility not to overstate their case and claim hypotheses as truth. Some knowledge is secure and provable, but much of our evolving science is not yet at this level.

We do not understand why the laws of nature are the way they are. We live in a universe which is finely tuned, and the question is *why*. Are the laws of nature *descriptive*, simply characterizing the way things are, or are they *purposeful*, forcing them to be this way?[1] Has the universe expanded at a most precise rate to allow for the existence of human beings—ourselves—fourteen billion years down the line? Enter the *anthropic principle*, which, according to cosmologist Brandon Carter, is a middle ground between the excessive anthropocentrism of the pre-Copernican age (an earth-centered cosmos) and the equally unjustifiable antithesis that no place or time in the universe can be privileged in any way. He elaborates as follows:

> In the form in which it was originally expounded, the anthropic principle was presented as a warning to astrophysical and cosmological theorists of the risk of error in the interpretation of astronomical and cosmological information unless due account is taken of the biological restraints under which the information was acquired. However, the converse message is also valid: biological theorists also run the risk of error in the interpretation of the evolutionary record unless they take due heed of the astrophysical restraints under which evolution took place.[2]

We do live at a privileged time; we are here, in a universe containing galaxies (see fig. 9), stars, and planets, reading this book! We do live in a privileged place—on Earth, orbiting the sun—and not within the bulges of galaxies, where the density of stars is very high and there may never be night to observe the universe. Such privileges and fine-tunings contain strong notions of design—of a Creator—which to some is utterly repugnant.

---

1. John W. Carroll, ed., *Readings on Laws of Nature* (Pittsburgh: University of Pittsburgh Press, 2004).
2. Brandon Carter, "The Anthropic Principle and Its Implications for Biological Evolution," *Philosophical Transactions of the Royal Society of London*, ser. A, *Mathematical and Physical Sciences* 310, no. 1512 (1983): 347.

There has been a fascinating response to this notion, which goes completely against the ground (of observability and testability) so firmly held by Galileo. This bold hypothesis has appeared on the "scientific" scene: that while our universe is indeed finely tuned, it is just one of a large ensemble or set of universes called *multiverses*. Proponents of the multiverse theory argue that while our universe may have the appearance of being privileged, there are myriad other universes; much like the analogy of "blowing bubbles"—each bubble representing a universe. We just happen to live in the "right bubble," they would argue. It is important for us to stress that these other universes are *not* observable, and we do *not* know whether they exist. The concept of multiverses is a *hypothesis*.

Cosmologist George Ellis elaborates:

Let me state it more strongly: it is dangerous to weaken the grounds of scientific proof in order to include multiverses under the mantle of "tested science." It is a *retrograde* step towards the claim that we can establish the nature of the universe by pure thought *without* having to confirm our theories by observational or experimental tests. This *abandons* the key principle that has led to the extraordinary success of science. The claim that multiverses exist *is a belief* rather than an established scientific fact.[3]

How would Galileo respond if he were alive today? He may well assert that his cherished pillar of observational proof has sunk deep into the oceans, that the integrity of science may be at stake with the claim of the existence of multiverses. We are not trying to downplay the value of hypotheses here. They are the route by which science progresses, but they must be recognized as such and not regarded as part of our collected base of secure and tested knowledge.

The book of nature has a range of related qualities, reflecting our very incomplete knowledge of it. The contents of this book range from established facts to half-formed theories and opinions.

---

3. George Ellis, "Opposing the Multiverse," *Astronomy and Geophysics* 49, no. 2 (2008): 2.5–2.7; italics added.

This is the way science works. Galileo entreats "these most prudent Fathers to consider very carefully the difference between statements which are a matter of opinion and those which can be demonstrated."

> *There is a great difference between commanding a mathematician or a philosopher and persuading a merchant or a lawyer to change their mind. It is not as easy to change one's view of conclusions which have been demonstrated in the natural world or in the heavens, as it is to change one's opinion on what is or is not permissible in a contract, a declaration of income, or a bill of exchange. The Church Fathers understood this very well. . . . In particular, we have the following words of St Augustine:*
>
>> *It is unquestionable that whatever the sages of this world have demonstrated concerning physical matters, we can show not to be contrary to our Scripture. But whatever they teach in their books that is contrary to Holy Scripture is without doubt wrong and, to the best of our ability, we should make this evident. And let us keep faith in our Lord, in whom are hidden all the treasures of wisdom, so that we will not be led astray by the glib talk of false philosophy or frightened by the superstition of counterfeit religion.*

According to Augustine, as quoted previously by Galileo, science is never at variance with the Bible if the science is correct and the Bible is correctly interpreted. We believe that the meaning of the Augustine quotation that Galileo included may be elaborated by another translation of it, as follows:

> It is to be held as an unquestionable truth that whatever the sages of this world have demonstrated concerning physical matters [can be] in no way contrary to our Bibles; if physical facts are clearly demonstrated and appear to be contrary to the Holy Scriptures, then *the interpretation of what is written in the Holy Scriptures* may be concluded without any hesitation to be quite false [the crucial word here being "interpretation"]. And according to our ability let us make this evident, and let us keep

the faith of our Lord, in whom are hidden all the treasures of wisdom, so that we neither become seduced by the verbiage of false philosophy nor frightened by the superstition of counterfeit religion.[4]

Galileo was trying to home in on the critical qualities that distinguish at least parts of the book of nature from the book of Scripture. He contrasted the constraints of the deliberations of mathematicians (based on firm external realities) with the man-made constraints of the lawyer or merchant.

There are statements that we can accept as firm experimental fact (like the existence of the Medicean moons of Jupiter; see fig. 10), and then there are others that are still matters of opinion (like the nature of the dark energy believed to be responsible for the acceleration of the expanding universe). Some of these statements have been passed down as part of the heritage of science (like Archimedes's principle), others have been established in more recent times (as with Galileo's discoveries), while yet others are in a formative stage and are not yet ready to be accepted as fact. Many of the latter will turn out to be incorrect and be discarded. In this context, the quote from Augustine may be aimed at the unestablished opinions of the sages rather than at firm facts that they have learned or observed.

Science is an evolving discipline in which lie many mysteries, masked from our eyes at the current time. An elementary example of one such mystery in astronomy: What *is* a galaxy? Can astronomers currently image a spiral galaxy at a telescope, showing the extent of its halo of dark matter? No. One of us (Block) penned these thoughts:

> What we see is only a representation of what actually is. The definition of a galaxy given in the Hubble Atlas beautifully portrays the treachery of photographic images of galaxies, being then defined as closed dynamical systems, or island universes.[5]

---

4. Augustine, quoted in Galileo, *Letter to the Grand Duchess Christina*, in *Discoveries and Opinions of Galileo*, trans. and ed. Stillman Drake (New York: Anchor Books, 1957), 194–95.

5. David Block, "Rings in Spiral Galaxies in the Local Group: Lessons from René Magritte," in *Lessons from the Local Group: A Conference in Honour of David Block and Bruce Elmegreen*, ed. Kenneth C. Freeman, Bruce G. Elmegreen, David L. Block, and Matthew Woolway (New York: Springer, 2015), 423–41.

The essential point here is that galaxies are invariably not isolated units in space, as one might at first glance glean, for example, from paging through the exquisite photographs made by the astronomer James Edward Keeler between 1898 and 1900. In the preface to that atlas, we read these words:

> Professor Keeler's photographs enabled him to make two discoveries of prime importance, not to mention several that are scarcely secondary to them. . . . [First,] many thousands of unrecorded nebulae [galaxies] exist in the sky . . . [and second,] . . . most of these nebulae have a spiral structure.[6]

Why do spiral galaxies have the spiral shapes they do, as seen in the early photographs by Keeler and others? Is there one coherent mathematical theory, accepted by all astronomers, to explain the theory of spiral structure? No!

Such is the magic of reality, to borrow some words from Richard Dawkins. However, as science continues on its path, there can be no contradiction between the two books if each is contextualized. To embrace both ensures that one keeps "the faith of our Lord" while freely exploring the book of nature with its grand number of masks.

That there was a great degree of arrogance among the theologians surrounding the Galileo affair cannot be denied, and a similar arrogance unfortunately appears now in the realm of science. Scientists in our time should, with humility, learn from the Galileo affair, lest they fall into the same trap of blindness. To quote Alister and Joanna McGrath,

> During the 1990s, Dawkins introduced the idea of God as some kind of a mental virus that infected otherwise healthy minds. It was a powerful image that appealed to a growing public awareness of the risk of physical infections from HIV and software infections from computer viruses.[7]

---

6. James Edward Keeler, "Photographs of Nebulae and Clusters, made with the Crossley Reflector," *Publications of the Lick Observatory, University of California Publications* 8 (1908): 7. Keeler died in August 1900, and the masterful 1908 photographic atlas appeared some years after his death.
7. Alister McGrath and Joanna Collicutt McGrath, *The Dawkins Delusion? Atheist Fundamentalism and the Denial of the Divine* (Downers Grove, IL: IVP Books, 2007), 68–69.

Pronouncements from famous public scientists can carry a lot of weight with those who are not themselves scientists. Atheist fundamentalist scientists use *scientific* jargon to whet the appetite of the public—their agenda is *very* subtle. *Viruses*—what an evocative word to use, with its flashing red *warning signs* everywhere—"Hazardous," "Danger," "Stay Away." The flaw here is that there is absolutely nothing *scientific* about such a *hypothesis*. God is a virus of the mind? There is no way that science can prove or disprove the existence of God. Of course it cannot; such are the boundaries of the book of nature.

Galileo understood this well:

Now if scientific conclusions which are demonstrated to be true should not be made subordinate to Scripture, but rather the text of Scripture should be shown not to be contrary to such conclusions, it follows that before a scientific statement is condemned it must be shown that it has not been conclusively demonstrated. And the responsibility for showing this must lie not with those who uphold its truth but with those who believe it to be false: this is only reasonable and natural, for it is much easier for those who do not believe a statement to identify its weaknesses than for those who believe it to be true and conclusive. Indeed, the upholders of an opinion will find that the more they go over the arguments, examining their logic, replicating their observations, and comparing their experiments, the more they will be confirmed in their belief.

While Galileo was absolutely correct that "demonstrated physical conclusions need not be subordinated to biblical passages," neither can spiritual truths (such as the existence of God and our salvation) be subordinated to scientists. Science is simply not in a position to pontificate on our relationship with the Logos, the God-man.

Galileo emphasized that the scientific method itself is based not on whim but on rigor, involving experiment and observations. Repetition of those observations leads to confirmation of hypotheses: this is what Galileo's opponents failed to embrace. Science needs to be falsified by using the scientific method, not by simply quoting the Scriptures. This is indeed the thrust of Galileo's entire letter to the Duchess, that it is the domain of scientists to verify or disprove scientific theories. It is not the place of theologians to falsify scientific ideas using bare scriptural arguments—even as it is appropriate for theologians to explore how actual scientific findings and Scripture ultimately harmonize. In Galileo's view, the responsibility for proving that a theory is false rests with scientists who think that the theory is incorrect. This is the manner in which science works today: scientific opponents of a theory have the motivation to find its flaws.

Killing the Light

> To ban Copernicus' book now, when many new observations and the work of many scholars who have read it are establishing the truth of his position and the soundness of his teaching more firmly every day, . . . would in my view seem to be a contravention of the truth. . . . Not to ban the whole book, but just to condemn this particular proposition [of a moving earth] as false, would, if I am not mistaken, be even more harmful to people's souls, for it would allow them to see the proof of a proposition which they were then told it was sinful to believe. And to forbid the whole science of astronomy would be nothing less than contradicting a hundred passages of Holy Scripture, which teach us that the glory and greatness of God is wonderfully revealed in all his works, and made known divinely in the open book of the heavens. Nor should anyone think that the lofty concepts which are to be found there end in simply seeing the splendour of the Sun and the stars in their rising and setting, which is as far as the eyes of brutes and the common people can see. The book of the heavens contains such profound mysteries and such sublime concepts that all the burning of midnight oil, all the labours, and all the studies undertaken by hundreds of the most acute minds have still not fully penetrated them, even after investigations which have continued for thousands of years.

How contrary censorship is to the open book of nature.

We are reminded of the speech against censorship by the English author John Milton (1608–1674) to Parliament in England. This speech is called the *Areopagitica* and is among history's most influential and impassioned philosophical defenses of the principle of a right to freedom of speech and expression. Written in opposition to licensing and censorship, the speech expressed many principles that form the basis for modern justifications of freedom of the press.

Jonathan Rosen recalls the story:

> Sometime in 1638, John Milton visited Galileo Galilei in Florence. The great astronomer was old and blind and under house arrest, confined by order of the Inquisition, which had forced him to recant his belief that the earth revolves around the sun, as formulated in his "Dialogue Concerning the Two Chief World Systems." Milton was thirty years old—his own blindness, his own arrest, and his own cosmological epic, "Paradise Lost," all lay before him. But the encounter left a deep imprint on him. It crept into "Paradise Lost," where Satan's shield looks like the moon seen through Galileo's telescope, and in Milton's great defense of free speech, "Areopagitica," Milton recalls his visit to Galileo and warns that England will buckle under inquisitorial forces if it bows to censorship, "an undeserved thraldom upon learning."[8]

We would stress that the book of nature can never be suppressed. The open book of the universe can be unraveled only by detailed examination and observation. Even today, through the penetrating eyes of the world's largest telescopes, the universe remains filled with mystery.

Galileo cautioned against the example of banning on theological grounds a particular scientific proposition, which may subsequently be proved correct. He saw this as potentially "detrimental to the minds of men" by bringing the church into disrepute and ridicule. We would go along with this view and its counterpart: proposing a theological explanation for some as-yet-not-understood scientific phenomenon, which may subsequently be explained on physical grounds alone. This

---

8. Jonathan Rosen, "Return to Paradise: The Enduring Relevance of John Milton," *New Yorker*, June 2, 2008, http://www.newyorker.com/magazine/2008/06/02/return-to-paradise.

is the "God of the gaps" scenario discussed earlier and is in the same sense "detrimental to the minds of men."

In reflecting on truth and the book of Copernicus, one is reminded again by Milton of the invaluable role of (uncensored) books in general. We give two excerpts from his famous *Areopagitica*. The first one reads thus:

> For books are not absolutely dead things, but do contain a potency of life in them to be as active as that soul whose progeny they are; nay, they do preserve as in a vial the purest efficacy and extraction of that living intellect that bred them.[9]

Milton continues:

> As good almost kill a man as kill a good book. Who kills a man kills a reasonable creature, God's image; but he who destroys a good book, kills reason itself, kills the image of God, as it were in the eye.[10]

Scientists like Blaise Pascal, who truly *see* with both their physical and spiritual eyes, dwell in the land of the light; his "night of fire" illuminated his being, removing all blindness. For Pascal, Jesus, who grants us salvation, is the jewel beyond all price; the cathedral of science—that is, scientism—never displaced the God of Pascal. The nature of truth lies in the stereoscopic view of the physical and spiritual realms. To insist that truth lies only in one or the other domain is tunnel vision, which belongs to the land of the blind.

We reaffirm the wise words from Augustine, cited by Galileo: "Let us keep the faith of our Lord, in whom are hidden all the treasures of wisdom."[11] The apostle Paul kept the faith of our Lord. He described the harmony of God controlling both the macrocosm (the universe) and the microcosm (our heart) so movingly in 2 Corinthians 4:6: "For God, who commanded the light to shine out of darkness, hath shined in our hearts, to give the light of the knowledge of the glory of God in the face of Jesus Christ" (KJV).

---

9. John Milton, *Areopagitica* (London: Adam and Charles Black, 1911), 8.
10. Milton, *Areopagitica*, 9.
11. Augustine, quoted in Galileo, *Discoveries and Opinions of Galileo*, 194.

Historical Excursus: The Trial of Galileo

What was the principal reason why Pope Urban VIII summoned Galileo to Rome and the Inquisition? We wonder if Urban VIII was seeing far ahead to how this whole controversy would play out for the church, how the church would lose much influence and power, and how theologians would no longer control the realm of the scientific process. Urban VIII may have recognized this event as the start of the Scientific Revolution, and perhaps he could see the consequences for the church that have evolved in the centuries between Galileo and our time. Was the science of astronomy the secondary issue and the power play from theologians the real issue? We would argue yes, as stated earlier: Urban VIII could live with treating the Copernican view as a hypothesis, but it was Galileo's insistence on pushing the Copernican hypothesis as fact that precipitated the confrontation. What could Urban VIII have done to avert this potential catastrophe? Did he think that Galileo's retraction would change anything? Did he think that this public example would be the end of the revolution? Looking back with a few centuries of hindsight, we can see the futility of trying to suppress the truth. Maybe this was not so apparent at the time.

Aging, ailing, and threatened with torture by the Inquisition, Galileo recanted on April 30, 1633. Galileo had pleaded ill health prior to traveling from Florence to Rome. The declaration Galileo sent had been signed by three doctors on December 17, 1632.

The response, after a meeting of the Holy Office in Rome on December 30, 1632, quickly followed:

> His Holiness [the pope] and the Sacred Congregation will send a commissary there [to Florence] together with doctors who will visit him and make a specific and clear report on his [Galileo's] state of health; and if he is in such a state as to be able to come [to Rome], they should send him *imprisoned and in chains*; if on the other hand, because of his health and the danger to his life it will be

> necessary to put off the move [to face the Inquisition in Rome], as soon as he will have recovered and the danger has ceased, he should be transported *imprisoned and in chains*. Furthermore, the Commissary and the doctors should be sent at his [Galileo's] expense.[12]

These were the days of the Inquisition. Chains if need be.

Galileo was severely ridiculed in a letter that Cardinal Antonio Barberini wrote, at the order of Pope Urban VIII, to the inquisitor in Florence:

> He [Galileo] should not excuse his disobedience [to come to Rome] because of sickness. . . . It is very bad for him to seek an excuse by *making out* he is sick, for the fact is that His Holiness [the pope] and these my most eminent Lords do not wish in any way to *tolerate these fictions*.[13]

We were privileged to view the actual Inquisition documents at the Vatican. This was an emotional experience. The trial highlighted blindness, bias, and prejudice on the part of the theologians to the scientific discoveries of Galileo; the tension between the two books had reached its pinnacle. The apparent chasm between the book of nature and the book of Scripture reached a climax at the trial of Galileo. Its rippling effects are still evident in multitudes of books denying the existence of God today. According to our Lord, one reaps what one sows.

The following timeline gives an overview of exactly what led Urban VIII to summon Galileo to Rome at his advanced age.[14] What had Galileo done, what books had he written, and who were some of the key figures against him?

---

12. Annibale Fantoli, *Galileo: For Copernicanism and the Church*, trans. George V. Coyne, Vatican Observatory Publications, Studi Galleiani 3 (Notre Dame, IN: University of Notre Dame Press, 1994), 294; italics added.

13. Fantoli, *Galileo*, 295; italics added.

14. Douglas O. Linder, "Trial of Galileo: A Chronology," Famous Trials, accessed June 26, 2018, http://www.famous-trials.com/galileotrial/1015-chronology. Reproduced by permission of the author.

November 1613	Dominican friar Niccolò Lorini of Florence, a professor of ecclesiastical history, launches an attack on Galileo.
December 1613	Galileo writes a letter to Benedetto Castelli, a professor of mathematics at the University of Pisa, offering his ideas concerning the relationship of science and Scripture.
February 1615	Father Lorini files a complaint with the Roman Inquisition against Galileo's Copernican views. Included with the complaint is a copy of Galileo's 1613 letter to Castelli.
April 1615	Cardinal Robert Bellarmine cautions scientists to treat Copernican views as hypothesis, not fact.
December 1615	Galileo travels to Rome to defend his Copernican views.
January 1616	Galileo argues in writing that tidal motion proves that the earth revolves.
February 1616	A committee of advisers to the Inquisition declares that holding the view that the sun is the center of the universe or that the earth moves is absurd and formally heretical.
February 26, 1616	Cardinal Bellarmine warns Galileo not to hold, teach, or defend Copernican theory. According to an

	unsigned transcript found in the Inquisition file in 1633, Galileo is also enjoined from discussing his theory, either orally or in writing.
March 1616	The Congregation of the Index bans Copernicus's *On the Revolutions of the Heavenly Spheres* until corrections can be added. Galileo meets with Pope Paul V.
1619	After three comets appear in 1618 and prompt widespread speculation as to their nature, Galileo writes his treatise *Discourse on Comets*, which disputes Jesuit views on the subject.
1621	Galileo is elected consul of the Accademia Fiorentina. Pope Paul V dies and is succeed by Pope Gregory XV.
1623	Pope Gregory XV dies. Cardinal Maffeo Barberini is named Pope Urban VIII. Galileo publishes *The Assayer*, which offers his explanation for sunspots and comets.
1624	Galileo goes to Rome. He has six audiences with the pope and meets with influential cardinals. Pope Urban VIII tells Galileo that he can discuss Copernican theory—provided he treats it as a hypothesis.
April 1630	Galileo completes work on his satirical *Dialogue concerning the Two Chief World Systems*, in

	which he strongly promotes the Copernican theory as fact and in which Pope Urban VIII can be identified as one of the characters in the dialogue.[15]
June 1630	Galileo obtains conditional approval from the secretary of the Vatican for publication of *Dialogue concerning the Two Chief World Systems*.
February 1632	*Dialogue concerning the Two Chief World Systems* is printed.
Summer 1632	Distribution of *Dialogue concerning the Two Chief World Systems* is stopped by Pope Urban VIII. The pope authorizes a special commission to examine the book.
September 1632	Based on the special commission's report, the pope refers Galileo's case to the Inquisition.
October 1632	Galileo receives a summons to appear before the Inquisition. An aging Galileo requests that his trial be moved to Florence.
November 1632	Galileo's request to have his trial transferred to Florence is refused.
December 1632	Three physicians declare that Galileo is too ill to travel to Rome. The Inquisition rejects the

---

15. In hindsight, it seems that Galileo lacked some wisdom at this point. We wonder, in fact, what were the long-term agendas of both Galileo and Urban VIII in this encounter? What did Galileo think he would achieve by provoking Urban VIII so sorely in this popular-level book? And what did Urban VIII think he would achieve by suppressing it?

	physicians' statement and declares that if Galileo does not travel to Rome voluntarily, he will be arrested and taken in chains.
February 1633	Galileo arrives in Rome. He is allowed to stay at the home of the Tuscan ambassador but is forbidden to have social contacts.
April 1633	Galileo is interrogated before the Inquisition. For over two weeks, he is imprisoned in an apartment in the Inquisition building. Galileo agrees to plead guilty to a lesser charge in exchange for a more lenient sentence. He declares that the Copernican case was made too strongly in his book *Dialogue concerning the Two Chief World Systems*.
June 22, 1633	Galileo is sentenced to prison for an indefinite term.[16] Seven of ten cardinals presiding at his trial sign the sentencing order. Galileo signs a formal recantation. Galileo is allowed to serve his term under house arrest in the home of the archbishop of Siena.
December 1633	Galileo is allowed to return to his villa in Florence, where he lives under house arrest.

---

16. For the full text of Galileo's sentence, see Giorgio de Santillana, *The Crime of Galileo* (Chicago: University of Chicago Press, 1955), 306–10. Also available online at Douglas O. Linder, "Papal Condemnation (Sentence) of Galileo," Famous Trials, accessed September 13, 2018, http://www.famous-trials.com/galileotrial/1012-condemnation.

April 1634	Galileo's daughter, Maria Celeste, dies.
1638	Galileo is now totally blind. He petitions the Inquisition to be freed, but his petition is denied. John Milton visits Galileo.
1641	In his last major contribution, Galileo proposes using pendulums in clocks.
January 8, 1642	Galileo dies in Arcetri in his home villa Il Gioiello, "the Jewel" (see fig. 11), situated on the Arcetri hill in the village of Pian dei Giullari.
1820	The papal Inquisition is abolished.
September 11, 1822	The college of cardinals announces that "the printing and publication of works treating of the motion of the earth and the stability of the sun, in accordance with the opinion of modern astronomers, is permitted." Two weeks later, Pope Pius VII ratifies the cardinals' decree.
1835	Galileo's *Dialogue concerning the Two Chief World Systems* is removed from the Vatican's list of banned books.
1992	The Catholic Church formally admits that Galileo's heliocentric views, that the planets orbit the sun, are correct.

The trial of Galileo was one of the most dramatic events in the history of science. After the reading of the sentence at his trial, Galileo knelt to recite his abjuration, a portion of which reads,

> Therefore, desiring to remove from the minds of your Eminences and every faithful Christian *this vehement suspicion*, rightly conceived against me, with a sincere heart and unfeigned faith *I abjure, curse, and detest the above mentioned errors and heresies*, and in general each and every other error, heresy, and sect contrary to the Holy Church; and I swear that in the future I will never again say or assert, orally or in writing, anything which might cause a similar suspicion about me....
>
> I, the above-mentioned Galileo Galilei, have abjured, sworn, promised, and obliged myself as above; and in witness of the truth I have signed with my own hand the present document of abjuration and have recited it word for word in Rome, at the convent of the Minerva, this twenty-second day of June 1633.
>
> I, Galileo Galilei, have abjured as above, by my own hand.[17]

The two of us were able to examine the handwritten document of abjuration at the Vatican, signed by Galileo himself; we cannot imagine the deepest of emotions that Galileo must have felt as he signed the Inquisition documents. It was a betrayal of everything that he stood for. But despite his abjuration, this was such a pivotal time in the history of science: while the father of modern science, Galileo himself, was sentenced to house arrest in Florence, the foundations of modern science had truly been laid.

Others during Galileo's lifetime could see the approaching conflict between the emerging science and long-held beliefs. One of the words for the newly invented telescope or spyglass

---

17. Quoted in Maurice A. Finocchiaro, trans. and ed., *The Trial of Galileo: Essential Documents* (Indianapolis: Hackett, 2014), 138–39; italics added.

was *cannone* ("tube"). The Benedictine monk Angelo Grillo (ca. 1560–1629) could see where usage of the new instrument was leading. In the words of historians of science Massimo Bucciantini, Michele Camerota, and Franco Giudice, "Grillo underscored all the audacity of this new instrument with a shudder: 'In this sense, every man can be called a *cannonista*.'"[18]

Grillo imagined a person who uses a telescope as the user of an intellectual weapon. Anyone with a telescope could, in Grillo's view, *destroy* cherished and preconceived theological notions, such as the earth-centered universe. Metaphorically, Grillo saw those with a telescope as having the power *in their own hands* to "gun down" the power of the theologians.

Bucciantini, Camerota, and Giudice elaborate:

> Spare as they are, these words [of Grillo] portend troubling scenarios. The idea that "Galileo's large spyglass" would offer everyone and anyone the chance to see a new sky could have unexpected effects and, like a river in full spate, destroy every idea of order and hierarchy of fields of knowledge.[19]

---

18. Massimo Bucciantini, Michele Camerota, and Franco Giudice, *Galileo's Telescope: A European Story*, trans. Catherine Bolton (Cambridge, MA: Harvard University Press, 2015), 240.

19. Bucciantini, Camerota, and Giudice, *Galileo's Telescope*, 240.

6

An Illusion of Conflict

Sunrises and Sunsets

While science is a rational and intellectual process, the outcome is not always consistent with everyday experience. The scriptural view held by Galileo's opponents of a hierarchical cosmos with the stability of the earth and the sun rising and setting was indeed consistent with sunrise and sunset each day, but that preconceived structure of an earth-centered universe was incorrect. The correct scenario, of the rotating earth, is equally consistent with the observations of sunrise and sunset.

Galileo understood that there was no reason to suppose a conflict between the teaching of Scripture and his scientific discoveries. He wrote that the duty of believers in both grace and science was to seek the "the true meaning of Scripture in those places where it appears to state the opposite":

> From this and from other passages it seems to me, if I am not mistaken, that the view of the Church Fathers was that on questions of natural science which are not matters of faith, we should first consider whether they have been demonstrated beyond doubt or are known from the evidence of the senses, or whether such certain knowledge is possible. If it is, and since this too is a gift of God, we should apply ourselves to understanding the true meaning of Scripture in those places where it appears to state the opposite. Wise

> *theologians will undoubtedly be able to penetrate its true meaning, together with the reasons why the Holy Spirit should sometimes have chosen to veil it under words signifying something different, either to test us or for some other reason which is hidden from me.*

There is an interesting paradox here, of course, in that theologians of the day and the common people could see the sun rising and the sun setting, apparently in orbit around the earth, but Galileo's telescopes appeared to indicate otherwise. So one can understand why both theologians and the common people were not very sympathetic to Galileo upholding the sun-centered Copernican universe, which seemed so counter to common experience. But as we now know, the sun-centered solar system is correct, and we must understand Scripture in the light of scientific reality.

The misunderstanding Galileo encountered was widespread among the common laypeople. In fact, one of the reasons for the intense opposition to Galileo was just how much an earth-centered universe reigned in popular opinion. Galileo realized this, writing,

> *If you were to quiz a thousand men among the common people about their view of this matter, I doubt whether you would find one who did not declare himself firmly convinced that the Sun moves and the Earth stands still. But no one should take this almost universal popular consent as an argument for the truth of what they assert; for if we were to question these same men about their grounds and reasons for believing as they do, and on the other hand to listen to the experiments and proofs which have led a few others to believe the opposite, we would find that the latter are persuaded by solidly based reasons, while the former are influenced by shallow appearances and vain and ridiculous comparisons.*

A poignant lesson here is that popular consent is no ground for the truth. One must rather be consistent and follow the spirit of the book of Scripture: one must not say that something is true if it is clearly false, or false if it is clearly true, simply for the case of popular consent. There is no case for believing a lie: the whole spirit of Scripture is to assert the truth.

A Crisis of Power

Galileo spoke of the masses—quizzing one thousand men—and history brims with episodes of *power* and *control* of the masses. The preservation of control and belief must surely have deeply concerned Pope Urban VIII at the trial of Galileo: stretching wide on the horizon was the looming problem of how to integrate the new science into common experience and belief. Perhaps it was already too late. In the West, the ascendancy swung abruptly from the church in Galileo's time to science, even within the same century. In the times of great names such as John Newton, Gottfried Wilhelm Leibniz, Blaise Pascal, and Antonie van Leeuwenhoek, knowledge of mathematical and other scientific domains was exploding. One wonders what would have happened to the evolution of science versus religion if the concept of a sun-centered world had already flourished in the West during the times of early Christian fathers such as Augustine.

This could easily have happened. Aristarchus of Samos proposed a heliocentric solar system in the third century BC, but the idea did not catch on. It was not that there were no astronomical instruments prior to the time of Galileo but that all the instruments available required observations by eye, without the aid of a telescope. For example, Hipparchus (ca. 190–120 BC), who is regarded as the greatest astronomical observer of antiquity, used basic instruments such as the astrolabe for his astronomical observations. It is very appropriate that the first space experiment devoted to precision astrometry (the accurate measurement of the positions of celestial objects on the sky), launched in 1989 and operated by the European Space Agency until 1993, carried the name Hipparcos. (The word *Hipparcos* is an acronym for *high precision parallax collecting satellite*).

Computers are often regarded as a contemporary invention, but an ancient analog computer, the Antikythera mechanism (designed to predict astronomical positions and eclipses for calendar setting, astrology, and the cycles of the ancient Olympic Games) is believed to have been designed and constructed by Greek scientists prior to the birth of Jesus. The instrument has been dated either to between 150 and 100 BC, or, according to a more recent view, to an even earlier epoch (205 BC). Astronomy had also flourished at early times in

China. Naked-eye instruments like the armillary sphere may have been developed already by the fourth century BC.

In the West, the slow growth of science until the time of Galileo meant that the scientific "enlightenment," when it came, overwhelmed the religious basis of daily life in a way that may not have been so fast and extreme if scientific knowledge had already been more advanced in his time. We again ponder: What would have unfolded in the science and religion encounter in Galileo's time had the proposal of a heliocentric universe been accepted by the time of Augustine? We suspect that the Galileo crisis was inevitable. It was not really a crisis of theology and knowledge; it was basically a crisis of church power, as Pope Urban VIII no doubt saw most clearly.

Incomprehension and the Masses

The notion of a conflict between the book of Scripture and Galileo's telescopic discoveries was a powerful illusion. For his part, Galileo himself tended to lay excessive blame at the intellectual capacities of ordinary people. For example, Galileo unnecessarily attributed Scripture's mysterious language to an intentional "dumbing down" of the biblical message to accommodate the mental deficiencies of the masses:

> *It is clear, then, that it was necessary to attribute movement to the Sun and rest to the Earth so as not to confuse the limited understanding of the masses, making them stubborn and reluctant to believe in the principal articles which are absolutely matters of faith; and if this was necessary, then it is not surprising that it was done, with great prudence, in Holy Scripture. I would go further, and say that it was not only consideration for the incomprehension of the masses but the prevailing opinion at that time which led the scriptural writers to accommodate themselves, in matters not necessary to salvation, more to received opinion than to the essential truth of the matter.*

Galileo repeatedly shared a rather dim view of the common people and thus spoke of the dumbing down of Scriptures. In stark contrast, we must remember that Jesus selected very ordinary (common) people to be his disciples, from Matthew the tax collector to Peter the fisherman. Jesus himself said in Matthew 11:25–27,

> I thank thee, O Father, Lord of heaven and earth, because thou hast *hid these things* from the *wise and prudent,* and hast *revealed them unto babes.* Even so, Father: for so it seemed good in thy sight. All things are delivered unto me of my Father: and no man *knoweth* the Son, but the Father; neither *knoweth* any man the Father, save the Son, and he to whomsoever the Son will *reveal* him. (KJV)

There is intellectual discernment, and there is spiritual discernment. The two should not be confused. The Lord of heaven and earth *hides* the secrets of salvation from those wise in their own eyes but *reveals* them unto *babes.* Jesus *reveals himself* to the meek, the humble, by grace.

Galileo missed the mark here. Discipleship has nothing do with intellectual capacity or the "incomprehension of the masses." There is no watering down of Scripture; God loves all his creation (John 3:16). In 1 Corinthians 3:18–19, one finds a poignant contrast between our intellectual eyes and our spiritual eyes:

> Let no one deceive himself. If anyone among you thinks that he is wise in this age, let him become a fool so that he may become [truly] wise. For the wisdom of this world is folly with God. For it is written [in Scripture], "He catches the wise in their [own] craftiness." (ESV)

To make the Scriptures accessible only to a select group of trained theologians is against the entire spirit of the Gospels. Of course, there is an interesting interplay here—of the "common people" and the church of Rome, especially if the "common people" at that time relied almost exclusively on the church of Rome for biblical exegesis and for enlightenment and instruction regarding personal salvation. But this would only reinforce our criticism of Galileo's view that those in power should feed the people only what they can *intellectually* discern.

We feel that Galileo's view here is somewhat out of order. He argued that the allusions to physical phenomena in the Bible were written to accommodate the beliefs and comprehension of ordinary people. Why would he do so? With the looming risk of facing the Inquisition, he certainly had *much* motivation to avoid the fate of other perceived heretics. Galileo's pen resonated with a signal of danger in a letter to Giovanni Diodati (1576–1649) dated January 15, 1633:

> I am about to go to Rome, summoned by the Holy Office, which has already suspended my *Dialogue [concerning the Two Chief World Systems* (1632)]. From reliable sources I hear the Jesuit Fathers have managed to convince some very important persons that my book is execrable and more harmful to the Holy Church *than the writings of Luther and Calvin.*[1]

It is reasonable to suppose that Galileo spoke of accommodation in an effort to avoid speculations about biblical interpretation, because doing so would have been viewed as meddling in the territory of theologians. Appealing to accommodation was perhaps his way of dodging their wrath.

That said, we actually find Galileo here to be somewhat arrogant. Maurice Finocchiaro seems to concur: "Galileo was a master of wit and sarcasm."[2] Galileo's viewpoint was that the scribes watered down sections of Scripture for the sake of the common people. This is certainly not the line taken on spiritual issues, on which the Bible is forthright and uncompromising. We feel that Galileo's approach here is not the best way to understand the physically incomplete allusions to natural phenomena in the Bible. He argued that these allusions were written in order to be consistent with what the ordinary people could believe and comprehend. We would affirm that the central purpose of the Bible concerns our personal relationship with God and must be understood by all in their mother tongue. Thus, the basic message of the Bible, as captured in the gospel, is just as accessible

---

1. Maurice A. Finocchiaro, *The Galileo Affair: A Documentary History*, California Studies in the History of Science (Berkeley: University of California Press, 1989), 225; italics added. The scholar Giovanni Diodati first translated the Bible into Italian from the Hebrew and Greek. Born to a refugee Protestant family from Lucca in Italy in 1576, Diodati became a professor of theology in 1609. A leader of the Reformers, he was characterized by eloquence and boldness. Diodati was a fearless preacher.
2. Maurice A. Finocchiaro, trans. and ed., *The Trial of Galileo: Essential Documents* (Indianapolis: Hackett, 2014), 6.

to shepherds in the field as to highly trained theologians. It might be more profitable to ask why these allusions to natural phenomena appear in the Bible at all. We would argue that these physical allusions in the Bible are there as strings between the book of nature and the book of Scripture. They show how God is involved in the book of nature, as Creator of the physical universe. Had the book of Scripture been interpreted in this way at the time, the trial of Galileo may not have taken place. The theologians took these physical allusions in the Bible as *complete* descriptions of the cosmos. The apparent motion of the sun and the stationary nature of the earth were interpreted literalistically; with hindsight, we see that the theologians and church fathers were just as confused as the common people. But at the end of the day, was the key issue at stake the science, or was it power and absolute control? On which hinges did the door swing?

We had a most informative discussion on this issue with the distinguished Old Testament scholar Walter C. Kaiser Jr. on April 7, 2016. Kaiser stressed that just as science has a language, so, too, does the book of Scripture. Context is all important. When reading the Psalms, for example, poetic language is the correct context of interpretation. In Psalm 18:1 we find the phrase "The LORD is my rock" (ESV). The image is not of God as a physical rock; it should be understood in terms of his trustworthiness and his faithfulness, both of which are unshakable. Rather than Scripture being "dumbed down" for understanding by ordinary human beings untrained in theology, each verse resonates with a deep truth when interpreted in its context, with the symbolism or metaphor carefully understood.

Equally in the New Testament, this is all important: in John 10:9, for instance, Jesus says, "I am the door: by me if any man enter in, he shall be saved, and shall go in and out, and find pasture" (KJV). Our Lord is not for one moment implying that he is a physical door. Rather, by his Spirit, we have our spiritual eyes opened through him alone (who symbolically is "the door"), and we then enter a new dimension in life: that of grace and spiritual revelation.

In another passage, Jesus states, "I am the bread of life" (John 6:48 ESV). Jesus is not for one millisecond saying that we will find him by going into a nearby grocery store to buy bread. Jesus is saying something *literal* using *figurative* speech. He is the very essence of life, both now and beyond the grave—to the final resurrection of our bodies. In Philippians 3:10–11 the apostle Paul emphasizes this: "That I may know him, and the power of his resurrection, and the fellowship of his sufferings, being made conformable unto his death; If by any means I might attain unto the resurrection of the dead" (KJV).

In yet another illustration, in Psalm 104:5, it speaks of God, "Who laid the foundations of the earth, that it should not be removed for ever" (KJV). A literal interpretation would make no more sense than God being a rock or a door. The focus of the psalmist here is first on God and second on him meticulously establishing the earth in its path. The orbit cannot be moved. Astronomers recognize that the earth's orbit lies in a very narrow permissible zone—known as the habitable zone—which allows for the existence of each one of us as self-aware human beings. The earth's temperatures, atmospheric pressures, atmospheric content, magnetic force fields—and much besides—all reside in the "Goldilocks zone" for life. Likewise, Psalm 93:1, which says, "The world also is stablished, that it cannot be moved" (KJV), is steeped in the poetic tongue. Our world is firm; the laws of nature cannot be changed. The psalmist is actually painting a far deeper truth.

Kaiser emphasized the majestic and figurative image of God "holding up the earth"—his focus is on human beings and on their salvation. A crucial point that the opponents of Galileo had missed was that a literal interpretation of such verses in the Psalms would be greatly in error. One cannot force Scripture into a human mold that God never intended. It is not for mankind to prescribe the geography of the heavens on misguided theological grounds.

Kaiser quickly affirmed that there is no "dumbing down" here of Scripture but rather that poetic language may as much be inspired by God as a direct message introduced by the words "And God said." Certain sections of the book of Scripture are poetic, while others are literal, such as the battle of Jericho or David slaying the giant Goliath or God speaking to Moses or God dwelling among us in human form.

To finite human beings limited in time and space, figurative language is often highly effective and easily understood. We are thinking here specifically of the multitudes of parables used by our Lord, which do not require theological training to be comprehended and personalized by each one hearing them. They are intelligible to the common man, but their depth strains the intellectual resources of even the most highly educated theologians.

Such is the power of the book of Scripture: to probe the hearts of both cardinal and common man alike—no "dumbing down" required.

The Mysterious Language of Perception

Habit and tradition versus reality—bodies may visually appear to move one way, yet is that a depiction of reality? In his letter, Galileo gave some technical examples of how our impressions of what is happening in the nearby cosmos are not what is actually happening:

> *Copernicus himself recognized how much our imagination is influenced by ingrained habit and by ways of conceiving things which have been familiar to us since childhood; so in order not to make these abstract ideas even more confusing and difficult for us, once he had demonstrated that the movements which appear to us to belong to the Sun and the firmament are actually movements of the Earth, he continued to call them movements of the Sun and the heavens when he came to set them down in tables and show how they worked in practice. . . . This shows how natural it is to adapt ourselves to our habitual way of seeing things.*

The way things appear can be masked, as we see in the work of the Belgian surrealist René Magritte. His painting *The Treachery of Images* (*La trahison des images*) shows a pipe. Below it, Magritte painted, *Ceci n'est pas une pipe* ("This is not a pipe").

The way a pipe appears on a sheet of paper can be very different from a *real* pipe. Michel Foucault wrote a brilliant essay titled *Ceci n'est une pipe*, inspired by Magritte's painting. Foucault (quoting Magritte) notes, "It is in vain that we say what we see; what we see never resides in what we say. And it is in vain that we attempt to show, by the use of images, metaphors, or similes, what we are saying."[3]

---

3. Michel Foucault, *This Is Not a Pipe* [*Ceci n'est une pipe*], trans. and ed. James Harkness (Los Angeles: University of California Press, 1982), 9.

Foucault is affirming that we cannot describe the full compass of truth by sight alone; images are necessary but incomplete. The painting is indeed not a pipe but rather a representation of a pipe, which was Magritte's point. When Magritte was asked about this image, he replied that, of course, it was not a pipe—just try to fill it with tobacco!

The relationship between *language* used to describe the pipe and the painting of the pipe is complex—it is no one-to-one correspondence.[4] In a similar vein, the way the sun appears to move from east to west, as drawn by an artist on canvas, can stand in stark contrast to the truth. Underneath such a painting one can equally well write, "This is not the movement of the sun."

Astronomers may look at an optical photograph of our closest spiral galaxy, the Andromeda spiral, and say, "This is indeed the quiescent Andromeda spiral galaxy," but such is the treachery of images. When one looks at it with infrared eyes (see fig. 12), the infrared image one sees (and the *violent* history that the infrared image betrays) can be astonishing. One of us (Block) along with the fellow collaborators on the image in figure 12 explains:

> The unusual morphology of the Andromeda galaxy (Messier 31, the closest spiral galaxy to the Milky Way) has long been an enigma. Although regarded for decades as showing little evidence of a violent history, M31 has a well-known outer ring of star formation . . . whose centre is offset from the galaxy nucleus. In addition, the outer galaxy disk is warped, as seen at both optical and radio wavelengths. The halo contains numerous loops and ripples. Here we report the presence of a second, inner dust ring . . . and offset from the centre of the galaxy. The two rings appear to be density waves propagating in the disk. Numerical simulations indicate that both rings result from a companion galaxy plunging through the centre of the disk of M31. The most likely interloper is M32. Head-on collisions between galaxies are rare, but it appears nonetheless that one took place 210 million years ago in our Local Group of galaxies.[5]

---

4. Much further detail appears in David Block, "Rings in Spiral Galaxies in the Local Group: Lessons from René Magritte," in *Lessons from the Local Group: A Conference in Honour of David Block and Bruce Elmegreen*, ed. Kenneth C. Freeman, Bruce G. Elmegreen, David L. Block, and Matthew Woolway (New York: Springer, 2015), 423–41.

5. D. L. Block, F. Bournaud, F. Combes, R. Groess, P. Barmby, M. L. N. Ashby, G. G. Fazio, M. A. Pahre, and S. P. Willner, "An Almost Head-On Collision as the Origin of Two Off-Centre Rings in the Andromeda Galaxy," *Nature* 443, no. 7113 (2006): 832.

Perspective is crucial in the interpretation of imagery. Magritte masterfully emphasizes this point in his painting titled *La condition humaine*, or *The Human Condition*. Magritte writes,

> In front of a window seen from inside a room, I placed a painting representing exactly that portion of the landscape covered by the painting. Thus, the tree in the picture hid the tree behind it, outside the room. For the spectator, it was both inside the room and outside the real landscape.

To quote Galileo, it is important "not to make these abstract ideas even more confusing and difficult for us."

When interpreting imagery in the book of Scripture, those who opposed Galileo so vehemently were sadly lacking in perspective. It was "the treachery of images." Some preferred to remain *inside* their rooms, refusing to even look *outside* through a telescope, as Galileo affirmed in his letter to Kepler. To force Scripture to say what it does not say is to lose all perspective. The book of nature is an open book—the tree in Magritte's painting cannot be constrained in its geographical locale.

Galileo also referred to the power "exerted by custom"—the spirit of the age holds sway. While in Copernicus's time, the anchor lay in both books, today the anchor has been placed in the book of nature only. It has become "custom" to set it there, as Lance Morrow notes:

> Sometime after the Enlightenment, science and religion came to a gentleman's agreement. Science was for the real world: machines, manufactured things, medicines, guns, moon rockets. Religion was for everything else, the immeasurable: morals, sacraments, poetry, insanity, death, and some residual forms of politics and statesmanship. Religion became, in both senses of the word, immaterial. Science and religion were apples and oranges. So the pact said: render unto apples the things that are Caesar's, and unto oranges the things that are God's. Just as the Maya kept two calendars, one profane and one priestly, so Western science and religion fell into two different conceptions of the universe, two different vocabularies.[6]

---

6. Lance Morrow, *Fishing in the Tiber: Essays* (New York: Henry Holt, 1988), 195.

Although these vocabularies have parted ways, we need them both to talk about the entire system of the world in which we find ourselves, material and spiritual. To discard either vocabulary is unbalanced, like cutting off one of our legs. It is not for any one linguist to assert that only one vocabulary exists.

In our present, post-Galileo era, we are living in a society in which the role of God and the spiritual are rapidly disappearing. In our view, we are living in a topsy-turvy land in which the validity of our two books has been reversed. We are reminded of a famous essay by G. K. Chesterton, "In Topsy-Turvy Land," inspired by a poster he saw in a shop on Fleet Street in London. The poster carried the question "Should Shop Assistants Marry?" Why? Because a materialistic world focuses on profit. If shop assistants should marry, maybe some of their time and energy would be focused on their family—clearly not optimal for profit. Chesterton commented, "They do not ask if the means [of having a job] is suited to the end [of a family life]; they ask (with profound and penetrating skepticism) if the end [of a family life] is suited to the means."[7]

Which comes first: the human being or profit? Such is Chesterton's topsy-turvy land, where the priorities seem all out of balance—the shop assistant's human life does not enter the discussion. In his essay, Chesterton could see more broadly where the road of missing the focal point was leading to:

> The cross of St. Paul's might have been hanging in the air upside down. For I realize that I have really come into a topsy-turvy country; I have come into the country where men do definitely believe that the waving of the trees makes the wind. That is to say, they believe that the material circumstances, however black and twisted, are more important than the spiritual realities, however powerful and pure.[8]

Galileo's letter laments his own topsy-turvy situation, in which the church ignored what the heavens show us to hold onto faulty interpretations of Scripture passages—passages that could be interpreted in other ways that aligned with natural revelation.

---

7. G. K. Chesterton, "In Topsy-Turvy Land," in *Tremendous Trifles* (1909; repr., London: Methuen, 1930), 73.
8. Chesterton, "In Topsy-Turvy Land," 74.

Science Requires Faith Too

Galileo rejected the idea that a conflict between science and Scripture was unanimously perceived by all Catholics. He appealed to the work of an Augustinian hermit and academic who, in a commentary on the book of Job, appeared to endorse Copernicus's theories: Diego de Zúñiga of Salamanca, or Didacus à Stunica (1536–ca. 1597). But where would the church of Rome place one of the greatest astronomical treatises of all time—Copernicus's *De revolutionibus orbium coelestium*?

In a decree of the Sacred Congregation dated March 5, 1616 (shortly after Galileo had written his letter to the Duchess in 1615), we find the following words:

> This Holy Congregation has also learned about the spreading and acceptance by many of the false Pythagorean doctrine, altogether contrary to the Holy Scripture, that the earth moves and the sun is motionless, which is also taught by Nicolaus Copernicus's *On the Revolutions of the Heavenly Spheres* and by Diego de Zúñiga's *On Job*. . . . Therefore, *in order that this opinion may not advance any further* to the prejudice of Catholic truth, the Congregation has decided that the books by Nicolaus Copernicus (*On the Revolutions of the Heavenly Spheres*) and by Diego de Zúñiga (*On Job*) be suspended until corrected; but that the book of the Carmelite Father Paolo Antonio Foscarini be completely prohibited and condemned; and that *all other books which teach the same be likewise prohibited*, according to whether with the present decree it *prohibits*, *condemns*, and *suspends* them respectively.[9]

De Zúñiga correctly concluded that the mobility of the earth is not contrary to Scripture. Today, cosmologists studying our universe, on much larger scales than Galileo was considering, often invoke the cosmological principle. Astronomer William Keel explains this principle as follows:

---

9. Finocchiaro, *Trial of Galileo*, 103–4; italics added. The Carmelite Father Foscarini mentioned above had published a letter on the Pythagorean and Copernican ideas regarding the mobility of the earth and the stability of the sun. He was exploring ways to reconcile Scripture with these ideas. His letter did not go down well with Cardinal Bellarmine, who took the view that the Copernican ideas should be regarded as hypotheses and that reconciliation with the Bible was not permitted.

> The cosmological principle is usually stated formally as "Viewed on a sufficiently large scale, the properties of the universe are the same for all observers." This amounts to the strongly *philosophical* statement that the part of the universe which we can see is a fair sample, and that the same physical laws apply throughout. In essence, this in a sense says that the universe is knowable and is playing fair with scientists.[10]

The word "philosophical" is crucial. Long before Einstein's theory of relativity was invented, man was gradually being displaced from his position in the universe. First, we were led to believe by Copernicus that the earth was dethroned from its central position; next, that the sun was dethroned from any preferred locale; and finally, that we are immersed in the immensity of immensities, apparently living in a vast universe of zero purpose. Maybe we should see an agenda here. On the other hand, the cosmological principle is a mathematical device without which cosmologists would have a very hard time making progress.

Nevertheless, mankind is not dethroned. For many different reasons, we could not live in a universe that was much smaller (or much hotter). First, enough time is needed for the hot big-bang universe to cool off, for matter to form, and then for the matter and radiation to decouple. Next, we are carbon-based human beings. Carbon is manufactured deep in the interiors of stars. Galaxies must first form, then stars within those galaxies must be born and complete their life cycles; the end products of the more massive stars are the exploding supernovae. It is these explosions that unlock carbon and heavier elements from stellar interiors into space, from which new stars are formed. As best we can understand it, this process—from the birth of the universe to us being here, orbiting a star that is enriched in carbon—takes billions of years. The size of an expanding big-bang universe is inextricably connected to its age. A universe that has been expanding for some fourteen billion years is calculated to be about ninety-two billion light years in extent. For us to be here seems to require, from an astronomical perspective, an immense universe, spanning billions of light years. We should not be surprised to find ourselves living in a universe so large.

---

10. William C. Keel, *The Road to Galaxy Formation*, 2nd ed. (New York: Springer-Praxis, 2007), 2.

Galileo spoke about matters of faith. The notion that theologians—but not scientists—require faith is a myth. Scientists always require faith in their assumptions about the "unwritten sections" of the book of nature. One example is the force of gravity. Most scientists *believe* that gravity obeys an inverse square law; that is partly a statement of faith because it has not been well tested at very low accelerations. All we know about gravity is about the gravitational fields that can be measured to the outer limits of our solar system. One cannot travel billions of light years into space to test whether the inverse square law is valid throughout a universe spanning fourteen billion light years. Other scientists have contested that *belief*, claiming that in domains of very low accelerations (far lower than the acceleration of Pluto as it orbits the sun), Newton's law of gravity may need to be changed.

At the time of the church fathers, there was no apparent conflict between science and religion—spiritual mankind *was at the center of God's eye*. By the time of Galileo, the empirical methodology of science had only begun to appear. It was at this juncture that the church *enforced* their "molding of science" from the book of Scripture, which led to Galileo's *Letter to the Grand Duchess Christina of Tuscany*. The Grand Duchess had to be shown that the two books were still in harmony.

7

Discerning the Truth

Scripture and the Miracle of Incarnation

In hindsight, Galileo encountered a theological bias in those who looked on his views as heretical, with mind-sets that wrongly conflated ecclesiastical authority with biblical authority. Inquisitors had an ever-watchful eye for heretics, but the geography of the heavens cannot be molded by any member of the cloth. The universe with its billions of galaxies never yields to any ecclesiastical powers. In that light, Galileo turned his attention to the Council of Trent regarding matters of faith and doctrine:

> *I have, in any case, some reservations about the truth of the claim that the Church requires us to believe as articles of faith such conclusions in natural science as are supported solely by the common interpretation of the Church Fathers. I wonder whether those who argue in this way may have been tempted to extend the scope of the Conciliar decrees in support of their own opinion; for the only prohibition I can find on this matter is against distorting in a sense contrary to the teaching of the Church and the common consent of the Fathers those passages, and those alone, which concern matters of faith or morals or the building up of Christian doctrine. This is what was stated by the Council of Trent in its fourth Session.*

Described as the embodiment of the Counter-Reformation, the Council of Trent issued condemnations of what it defined to be heresies committed by Protestantism and key statements and clarifications of the doctrine and teachings of the Roman Catholic Church. This was in the period from 1545 to 1563, ending a year before Galileo was born, in 1564. It was a time when the Roman Catholic Church was under great pressure.

Much read by Protestants were the writings of the Venetian Paolo Sarpi (1552–1623), a contemporary of Galileo. Sarpi's greatest work is his *History of the Council of Trent* (*Istoria del Concilio Tridentino*). This monumental work established Sarpi as the most formidable adversary of the Counter-Reformation in Italy.

Why the Council of Trent? As noted by Frances Yates,

> In the earlier years of the sixteenth century many people were looking earnestly forward to a General Council of the Church in which points at issue between Catholics and Reformers should be resolved. . . .
>
> There were here involved two totally different conceptions as to the function of a General Council: one side held the view that the Protestants should be represented at it, and that a formula should be reached, under the guidance of the Holy Spirit, by which unity should be restored to the Church; the other side refused to consider concessions to Reformers and concentrated on the tightening up of an intense discipline under the Pope.[1]

This was not to be, and the Council of Trent became the charter of the Counter-Reformation. Yates notes,

> If the right course had been pursued at Trent, Sarpi indirectly suggests, the Church as a whole would have been reformed somewhat on the model of the Anglican reform (marriage of priests, Communion *sub utraque*, and the liturgy in the vernacular are all, of course, features of the Anglican Church). But the wrong course was pursued, and the Church, instead of being reformed, was deformed with new papal usurpations.[2]

David Wootton reflects as follows:

---

1. Frances A. Yates, "Paolo Sarpi's *History of the Council of Trent*," *Journal of the Warburg and Courtauld Institutes* 7 (1944): 132.
2. Yates, "Sarpi's *History*," 133.

Sarpi presents with care all the proposals for the reform of the Church and dissects the conflicting views of the theologians. . . . Above all, of course, he conveys the view that a true Council should be superior to the pope, that the Council should determine its own agenda rather than having one imposed upon it by the papal legates, and should not depend upon papal ratification for its decrees.[3]

Wootton notes that Sarpi viewed the Council of Trent as "a tragic history of hopes and expectations disappointed, of corruption and the abuse of power triumphant."[4]

Voltaire cited Sarpi's views of the Council of Trent as follows:

Finally we have the great Council of Trento,—but the dogma is indisputable, since the Holy Spirit used to come weekly from Rome to Trento, in the mail trunk, [according] to what Fra Paolo Sarpi says, however Fra Paolo Sarpi was slightly close to heresy.[5]

John Milton referred to Sarpi as the "great unmasker," while Wootton writes that the Catholic historian Hubert Jedin described Sarpi "as the papacy's greatest enemy after Luther."[6] His writings fueled intense anger and hatred, to the extent that assassins inflicted fifteen stiletto wounds on the body of Sarpi on October 5, 1607, and left him for dead, but he recovered. Alexander Robertson picks up the story:

[Sarpi] did not often refer to his enemies, but one or two utterances have come down to us. When the surgeon, Acquapendente, probing the most severe of the wounds, enlarged on its "stravaganza" (or roughness), Fra Paolo said, "E piore il monde vuole che sia

---

3. David Wootton, *Paolo Sarpi: Between Renaissance and Enlightenment* (Cambridge: Cambridge University Press, 1983), 106.
4. Wootton, *Paolo Sarpi*, 104.
5. Wootton discusses these views of Sarpi in *Paolo Sarpi*, 110. See also Voltaire, *Oeuvres complètes de Voltaire*, vol. 7, *Dictionnaire philosophique I* (Paris: Furne, 1847), 363. The original quotation in French reads, "Enfin nous avons le grand Concile de Trente,—mais le dogme en est incontestable, puisque le Saint-Esprit arrivait de Rome à Trente, toutes les semaines, dans la malle du courrier, à ce que dit Fra Paolo Sarpi, mais Fra Paolo Sarpi sentait un peu l'hérésie." We are grateful to Jacqueline Riffault-Silk and Francoise Combes for their English translations.
6. John Milton, *Areopagitica* (London: Adam and Charles Black, 1911), 16; Wootton, *Paolo Sarpi*, 120.

stata fatta 'stylo Romanœ Curiœ'" ["And yet the world says it was done in 'the style of the Roman Curia'"].[7]

There is, however, much more to the life of Sarpi. Of particular interest to us is that Paolo Sarpi was also an experimental scientist, a proponent of the Copernican system, and a one-time friend of Galileo. His role in the history and perfection of the telescope in the Republic of Venice should never be underestimated. We will meet Sarpi again in the pages to follow.

The Council of Trent determined that the church is the ultimate interpreter of Scripture and recognized the Bible *and the tradition of the church* as equally authoritative. In writing of "the office of grave and wise theologians to interpret the passages," Galileo was affirming the view of the church as the ultimate interpreter, and he urged integrity in interpreting Scripture. In contrast, we would place more emphasis on the role of the Holy Spirit in the personal revelation of God through the Scriptures and less emphasis on scriptural interpretation through "grave and wise theologians." Trent's decision on this issue would have serious implications for scientific matters.

We again emphasize here that during this period, Protestantism was not the only pressure bearing on the church of Rome. The Roman Catholic Church could see its power being eroded as the ultimate authority not only in the interpretation of the book of Scripture but also in the interpretation of the book of nature as the Scientific Revolution was starting to take hold. But Galileo reminds us of Augustine's view that the question "whether it is true that the heavens move or whether they are at rest" is secondary in biblical interpretation to the deeper issues of "salvation and . . . what is necessary and useful in the Church."

The primary focus of the book of Scripture is God's historical relationship with mankind. At the same time, as already noted, a string connects the book of Scripture to the book of nature. A fascinating thought is the thickness of the string. That string was very thick in a geocentric model of the universe. Has time frayed the string or even cut it?

---

7. Alexander Robertson, *Fra Paolo Sarpi: The Greatest of the Venetians* (London: George Allen, 1911), 186–87.

In a 1997 essay titled "Nonoverlapping Magisteria," the evolutionary biologist Stephen Jay Gould (1941–2002) placed a suggestion on the table: nonoverlapping magisteria (NOMA), which he developed to be "a blessedly simple and entirely conventional resolution to . . . the supposed conflict between science and religion."[8] Gould used the term *magisterium* from Pope Pius XII's 1950 encyclical *Humani generis*, defining it as "a domain where one form of teaching holds the appropriate tools for meaningful discourse and resolution," and he described the NOMA concept as follows:

> Science tries to document the factual character of the natural world, and to develop theories that coordinate and explain these facts. Religion, on the other hand, operates in the equally important, but utterly different, realm of human purposes, meanings, and values—subjects that the factual domain of science might illuminate, but can never resolve.[9]

Gould emphasized that "these two magisteria do not overlap, nor do they encompass all inquiry (consider, for example, the magisterium of art and the meaning of beauty)."[10] In the words of H. Allen Orr, however, "It is hard to resist the conclusion that Gould has lifted the word 'religion' and grafted it onto a toothless, hobbled beast incapable of scaring the materialists."[11]

We would argue that religion is very specifically about God's historical relation with mankind and mankind's response to his grace and revelation. This gives us our purpose and meaning. The values are not a set of rules that come from a mechanistic universe but are the fruits of God's Spirit, a consequence of our relationship with him.

For Christians, the incarnation of Jesus is a transformative step in this relationship and provides an unbreakable link between Gould's two magisteria, or what we are calling the books of nature and Scripture. God enters our physical world and opens up for us the path into salvation and God's kingdom. In G. K. Chesterton's poem,

---

8. Stephen Jay Gould, "Nonoverlapping Magisteria," *Natural History* 106, no. 2 (1997): 16–22.
9. Stephen Jay Gould, *Rocks of Ages: Science and Religion in the Fullness of Life* (New York: Ballantine Books, 2002), 4.
10. Gould, *Rocks of Ages*, 4.
11. H. Allen Orr, "Gould on God: Can Religion and Science Be Happily Reconciled?," *Boston Review*, October 1999, http://new.bostonreview.net/BR24.5/orr.html.

> There has fallen on earth for a token
> A god too great for the sky.[12]

In terms of John 1, describing God becoming flesh, the lifeline between the book of Scripture and the book of nature is intact, with spiritual man as the focus of God's creation. Incarnation is a miracle.

One viewpoint on miracles is naturalism (critiqued at length by Alvin Plantinga),[13] which claims that observable events are explained by natural causes only. Naturalism does not recognize phenomena commonly labeled *supernatural* (such as miracles).

In his book titled *Miracles*, C. S. Lewis carefully defines a miracle as "an interference with Nature by supernatural power," and he swiftly makes a distinction between two kinds of thinkers: the naturalist, who believes that nothing exists "except Nature," and the supernaturalist, who believes that "besides Nature, there exists something else."[14]

The mood toward miracles has largely turned toward skepticism, denial, and unbelief, resulting in part from the explanatory power of modern science beginning at the time of Galileo. Yet Lewis in his third chapter, "The Self-Contradiction of the Naturalist," argues,

> If my mental processes are determined wholly by the motion of atoms in my brain, I have no reason to suppose that my beliefs are true . . . and hence I have no reason for supposing my brain to be composed of atoms.
>
> To the naturalist, our cognitive faculties are selected for adaptation and for survival—not for truth. It follows that no account of the universe can be true unless that account leaves it possible for our thinking to be a real insight. A theory which explained everything else in the whole universe but which made it impossible to believe that our thinking was valid, would be utterly out of court. For that theory would itself have been reached by thinking, and if thinking is not valid that theory would, of course, be itself demolished. It would have destroyed its own credentials. It would be an argument which proved that no argument was

---

12. G. K. Chesterton, "Gloria in Profundis" (London: Faber & Gwyer, 1927). Public domain.
13. Alvin Plantinga, *Where the Conflict Really Lies: Science, Religion, and Naturalism* (New York: Oxford University Press, 2011).
14. C. S. Lewis, *Miracles: A Preliminary Study* (London: Geoffrey Bles, 1947), 15.

sound—a proof that there are no such things as proofs—which is nonsense.[15]

Philosopher Alvin Plantinga elaborates: "The naturalist can be reasonably sure that the neurophysiology underlying belief formation is *adaptive*, but nothing follows about the *truths* of the beliefs depending on that neurophysiology."[16]

We both accept the existence of miracles, in which God intervenes in the book of nature. We are not at all suggesting that the universe is a closed system in which God never intervenes. We would argue against such a view, for example, with the miracles of the incarnation and resurrection, the grace of answered prayer, and God's interventions in the daily lives of those who seek him.

Through the grace of personal and scriptural revelation, we both accept that God exists and is intimately involved in our lives. Salvation is a miracle. Blind or purposeless naturalism, therefore, makes no sense to us. Projecting the philosophy of naturalism into science is an agenda that we do not accept. Science itself provides no evidence that there is no God: it is the *philosophy of naturalism* that asserts that. We believe that the book of nature must be studied by the scientific method, not by the *philosophy of naturalism*, which has become a religion for some scientists today. It is not the place of science to go beyond its scientific bounds.

From a theistic point of view, we would expect our cognitive faculties to be reliable. God has created us in his image, and our ability to hold beliefs that are true is an important part of our image bearing. Jesus claimed that he is the truth, an absolute claim that is *not adaptive* to changes in our understanding of the universe from the time of Galileo to the present day. We are not naturalists, and we see it as perfectly consistent that God designed the early universe the way he did to prepare for the arrival of life at some epoch of its history and that God stepped into his universe some two thousand years ago. That is not all: the Gospels tell us that his redemptive plan for us is not yet

---

15. Lewis, *Miracles*, 28–29. Lewis is citing here the words of J. B. S. Haldane, *Possible Worlds and Other Essays* (London: Chatto and Windus, 1927), 209.
16. Alvin Plantinga, "The Dawkins Confusion," *Books & Culture*, March/April 2007, https://www.booksandculture.com/articles/2007/marapr/1.21.html.

complete. The book of Scripture tells us that *purpose* is enshrined in the creation of the entire universe, which includes mankind.

Benjamin Franklin and Chess

Benjamin Franklin (1706–1790) was one of the founding fathers of the United States. He was also a renowned scientist for his discoveries about electricity. Franklin had great insight as well into the game of chess. He penned these words in his 1786 essay "The Morals of Chess":

> The Game of Chess is not merely an idle amusement; several very valuable qualities of the mind, useful in the course of human life, are to be acquired and strengthened by it, so as to become habits ready on all occasions; for life is a kind of Chess, in which we have often points to gain, and competitors or adversaries to contend with, and in which there is a vast variety of good and ill events, that are, in some degree, the effect of prudence, or the want of it. By playing at Chess then, we may learn:
>
> I. Foresight, which looks a little into futurity, and considers the consequences that may attend an action. . . .
>
> II. Circumspection, which surveys the whole Chess-board, or scene of action:—the relation of the several Pieces, and their situations. . . .
>
> III. Caution, not to make our moves too hastily.[17]

In Galileo's time, a battle was raging, as in a game of chess, as to whether the earth was the center of the cosmos. More broadly, there was the issue of where to seek for information in the realm of nature—was it the Scriptures or the book of nature? Galileo asserted that this game was not being played by the ancient fathers, including Augustine. This was indeed an important question for Augustine among the ancient fathers, and he warned against theologians giving answers to questions for which they were ill equipped. But a different strategy was at play at the time of Galileo: many theologians wished

---

17. Benjamin Franklin, "The Morals of Chess," *Columbian Magazine* (1786): 159. For a discussion and excerpts, see John McCrary, "Chess and Benjamin Franklin: His Pioneering Contributions," http://www.benfranklin300.org/_etc_pdf/Chess_John_McCrary.pdf. See also Ralph K. Hagedorn, *Benjamin Franklin and Chess in Early America: A Review of the Literature* (Philadelphia: University of Pennsylvania Press, 1958).

to uphold their reputations by means of "vain fancies" devoid of experiment and observation.

Regrettably, many of Galileo's opponents exercised limited *caution* and *circumspection*: they failed to survey the *entire* arena, including the observations of Galileo made through his telescopes. And most regrettably, they failed to exercise foresight. Their foolish attempts to suppress physical truths quickly backfired, and the consequences are still felt today. This kind of suppression was doomed to failure as the Scientific Revolution took off, with the rapid transition from theology as queen to science as queen. With more foresight, the theologians of Galileo's day would not have put these books on their Index of forbidden books.

We also get a glimpse into the heart and mind of Galileo, who encouraged "a holy zeal for the truth, for Scripture, and for the majesty, dignity, and authority in which all Christians are bound to uphold it." We wonder how the transition from theology to science might have evolved if Galileo's opponents had more foresight and had acted more wisely.

As it is, the prosecution of Galileo was not a pursuit of truth as much as it was an opportunity for some people to flaunt their power over society. Galileo knew exactly what he was dealing with, and he spoke directly of the threatening power that the church flashed:

> *Surely it is plain to see that this dignity [of the Holy Scriptures] is far more zealously sought and secured by those who submit whole-heartedly to the Church, without asking for one or other opinion to be prohibited but only that they should be allowed to bring matters forward for discussion so that the Church can reach a decision [in matters scientific] with greater confidence, than by those who, blinded by their own self-interest or prompted by the malicious suggestions of others, preach that the Church should wield its sword straight away simply because it has the power to do so? Do they not realize that it is not always beneficial to do what one has power to do? This was not the view of the Church Fathers; on the contrary, they knew how prejudicial and how contrary to the primary intention of the Catholic Church it would be to use verses of Scripture to establish scientific conclusions which*

experience and necessary demonstrations might in time show to be contrary to the literal meaning of the text.

One is reminded of the grand inquisitor in the novel *The Brothers Karamazov* by Fyodor Dostoevsky, in which Jesus, again in human form, visits Spain during the Inquisition. Dostoevsky penned these words:

> The action of my poem takes place in Spain, at Seville, during the most terrible time of the Inquisition, when to the glory of God pyres were erected in the land and
>
> > "in the splendid *autos-da-fé*
> > Wicked heretics were burnt."
>
> ... In his boundless mercy he [Jesus] passes once again among the people in that very same human form in which he walked among the people fifteen centuries ago. He comes down in to the "hot lanes" of a southern city, in which city just the day before, "in a magnificent *auto-da-fé*," in the presence of the king, of the court, of the knights, of the cardinals and of the fairest ladies of the Court, in the presence of the large population of all Seville, had been burnt by the cardinal, the Grand Inquisitor, nearly a hundred heretics *ad majorem Dei gloriam*. He appeared quietly, imperceptibly, and—how strange—all the people recognize him. ... The people are drawn to him by an irresistible force, they surround him, flock round him, follow him. ... Lo, in the crowd, an old man, blind from childhood, cries out: "Lord, heal me that I may see Thee"; and scales, as it were, fall from his eyes and the blind man sees him.[18]

Dostoevsky continues by describing how Jesus is treated by the grand inquisitor in Seville. The read is a riveting one. The grand inquisitor informs Jesus that the church no longer needs him, as his return would disturb the mission of the church for power and control.

"Why then hast Thou come to hinder us?" asks the grand inquisitor.[19]

---

18. Fyodor M. Dostoevsky, *The Grand Inquisitor*, trans. S. S. Koteliansky (London: Elkin Mathews & Marrot, 1930), 5–6.
19. Dostoevsky, *Grand Inquisitor*, 11.

The novel nears its climax as the grand inquisitor says to Jesus, "I shall burn Thee for having come to hinder us. For if ever there was one who most of all deserved our fire, it is Thou. Tomorrow I shall burn Thee. *Dixi*."[20] This may appear extreme, but Dostoevsky was perplexed by the spirit of death and destruction at the time of the Spanish Inquisition, with its unspeakable cruelties and *autos-da-fé*.

There is a vehemence among certain atheists in detesting *religion*, and in one sense of the term, they are perfectly correct—looking at the Spanish Inquisition, to cite but one example. Ivan shares their abhorrence in his poignant portrayal of the clash between Jesus and the grand inquisitor[21]—a scene that arises, no doubt, from Dostoevsky's meticulous study of the Gospels.[22]

Religion and Jesus can be as far apart as the east is from the west. History is filled with untold horrors and cruelties in the name of organized religion with its man-made traditions. In contrast, there is the grace of God found in the Gospels.

Clergy themselves might vehemently oppose the nature of truth. It must be remembered that of the thousands of New Testaments translated by William Tyndale into English, almost all were met by *fire*. His translations were apparently heretical. In retrospect, it was Tyndale who "gave God an English voice," to quote David Teems.[23]

The true Jesus of the New Testament truly needed to be unmasked. In the annals of church history, he carries far too many masks of savagery and terror.[24] It is only through the lens of reading the New Testament with our own eyes and by the guidance of the Holy Spirit that we come to embrace the loving presence of "the luminous figure of the Nazarene" into our own hearts.

To whom might we now turn in our discourse but to John Milton, who, as noted earlier, met Galileo while he was living under house

---

20. By *Dixi*, he means, "I have said all that I have to say, and thus the argument is settled." Dostoevsky, *Grand Inquisitor*, 29.
21. Ivan (or Vanya) is "the dialectian" in the novel. He is subtle and pragmatic, and in this context, he represents the unbeliever.
22. Geir Kjetsaa, *Dostoevsky and His New Testament*, Slavica Norvegica 3 (Atlantic Highlands, NJ: Humanities Press, 1984).
23. David Teems, *Tyndale: The Man Who Gave God an English Voice* (Nashville: Thomas Nelson, 2012).
24. John Fox, *Fox's Book of Martyrs: The Acts and Monuments of the Church*, ed. John Cumming (London: George Virtue, 1844).

arrest in Florence. Milton alluded to this epoch of power and the flashing sword:

> When I recall to mind, at last, after so many dark ages, wherein the huge overshadowing train of error had almost swept all the stars out of the firmament of the church; how the bright and blissful Reformation, by divine power, struck through the black and settled night of ignorance and anti-Christian tyranny; methinks a sovereign and reviving joy must needs rush into the bosom of him who reads or hears, and the sweet odour of the returning Gospel, imbathe his soul with the fragrancy of heaven. Then was the sacred Bible sought out of the dusty corners, where profane falsehood and neglect had thrown it; the schools opened; divine and human learning raked out of the embers of forgotten tongues; the princes and cities trooping apace to the new erected banner of salvation; the martyrs with the unresistible might of weakness, shaking the powers of darkness, and scorning the fiery rage of the old red dragon.[25]

Milton spoke here in 1641, only a few years after Galileo's trial, of the glorious period of the Reformation, when the church's tyranny was receding and the two books were viewed with balance and proper regard. We wonder again if a wiser approach by Galileo's opponents could have enabled this equilibrium to be maintained. Their actions were driven by their perceived insecurity in dethroning the earth from its central locale to being merely one of many planets circling around the sun. Yet we are secure because of the grace of God, not because of orbits in space. Grace is found ultimately in the hand of God.

Galileo cautioned that theologians must not flash the sword of power to interpret the book of nature. Four hundred years down the line, we can say that the Bible never affirmed a geocentric worldview, but human ignorance imagined it. The hand of nature can never be forced into a model in which we as humans might feel more secure. Today it is not the theologians who flash their swords; science now has its own power-play agenda.

---

25. John Milton, quoted in George Offor, "Memoir of William Tyndale," in *The New Testament of Our Lord and Saviour Jesus Christ—Published in 1526—Being the First Translation from the Greek into English, by That Eminent Scholar and Martyr William Tyndale; With a Memoir of His Life and Writings, by George Offor* (London: Samuel Bagster, 1836), 4.

8

The Two Cathedrals

The Limits of Science

Speaking of Augustine, Galileo wrote,

> *Yet this author's caution is even more remarkable when, not being convinced after seeing the demonstrations, the literal meaning of Scripture, and the context of the passage as a whole all pointing to the same interpretation, he adds: "But if the context supplies nothing to disprove this to be the mind of the writer, we still have to enquire whether he may not have meant the other as well." . . .*
>
> *Finally, he justifies this rule of his by showing the dangers to which Scripture and the Church are exposed by those who, being more interested in maintaining their own error than in upholding the dignity of Scripture, seek to extend the authority of Scripture beyond the terms which Scripture itself prescribes. He adds the following words, which alone should suffice to restrain and moderate the excessive licence which some claim for themselves:*
>
>> It often happens that a non-Christian knows something about the earth, the heavens, and the other elements of this world, about the motion and orbit of the stars and even their size and relative positions, about the predictable eclipses of the Sun and Moon, the cycles of the years and the seasons, . . . and this knowledge he holds to as being certain

> *from reason and experience. Now, it is a disgraceful and dangerous thing for an infidel to hear a Christian, presumably giving the meaning of Holy Scripture, talking nonsense on these topics. We should take all means to prevent such an embarrassing situation in which the non-believer will scarce be able to contain his laughter seeing error written in the sky, as the proverb says. The shame is not so much that an ignorant individual is derided, but that people outside the household of the faith think our writers hold such opinions, and criticize and reject them as ignorant, to the great prejudice of those whose salvation we are seeking. When they find a Christian mistaken in a field which they themselves know well and hear him maintaining foolish opinions about our books, how are they going to believe those books in matters concerning the resurrection of the dead, the hope of eternal life, and the kingdom of heaven, when they think their pages are full of falsehoods about things which they themselves have learnt from experience and decisive argument?*

For theologians to argue from outside their trained discipline about issues that belong to the book of nature is to expose a crevasse that can bring our faith and our Bible into disrepute, "to the great prejudice of those whose salvation we are seeking." Of course, some theologians have applied themselves to advanced study of scientific issues and are better qualified to address such matters, but as Augustine describes here, too many speak confidently about science without ever carefully examining the book of nature.

As Augustine says, "We should take all means to prevent such an embarrassing situation in which the non-believer will scarce be able to contain his laughter seeing error written in the sky." Invoking God on biblical grounds as having designed a geocentric cosmos when Galileo's telescopic observations were to the contrary was one big step into this crevasse.

These were the traps for theologians at the time of Augustine and later in Galileo's day, when science was still undeveloped and theology and philosophy held sway. Today the pendulum has swung: science is ascendant, but there are still crevasses into which scientists can fall. Some scientists would claim that science has shown that God is not needed. Can science prove or disprove the existence of God? We have argued that it is simply impossible for science to disprove the existence of God because that is not a scientific question. To inject this agenda into the book of nature would be such a pitfall.

The chemist Peter Atkins believes that

> scientists, with their implicit trust in reductionism, are privileged to be at the summit of knowledge, and to see further into truth than any of their contemporaries. . . . There is no reason to expect that science cannot deal with *any* aspect of existence. . . . Science has never encountered a barrier that it has not surmounted or that we can at least reasonably suppose it has the power to surmount. . . . I do not consider that there is any corner of the real universe or the mental universe that is shielded from [science's] glare.[1]

These words remind us of a prevalent mood today: the *glare* of science, as Atkins terms it. Many people are today *transfixed* by science and science alone—just as a person in the African bush might be *transfixed* by the sight of a nearby leopard drinking water or perched in a nearby tree. Are we transfixed by the glare of science, as described above? Do the rocky crags echo with these diabolical words, "Science, behold your King?"[2] We assert that this is not the nature of truth at all.

For us, as two research astronomers who have devoted their careers to the study of galaxies, we know that science—the truth of nature—is still in the foothills of knowledge, and science's glare shines dimly into the corners of our subject. We are overwhelmed by known unknowns.

Let us give a few examples:

- We simply do not know what galaxies are actually made of: we know that 97 percent of their mass is in the form of mysterious dark matter, but we still have no idea what this dark matter is.

---

1. Peter W. Atkins, "The Limitless Power of Science," in *Nature's Imagination: The Frontiers of Scientific Vision*, ed. John Cornwell (New York: Oxford University Press, 1995), 125; italics added.
2. This is an allusion to John 19:14: "And it was the preparation of the passover, and about the sixth hour: and he [Pilate] saith unto the Jews, Behold your King!" (KJV). Is the shoe now completely on the other foot? Has the truth of nature (science) nullified the nature of truth?

- Why do spiral galaxies have their spiral shapes? We have theories, but none explains the complex variety of shapes that we see.
- Most galaxies have a supermassive black hole at their center. The Milky Way's central black hole has a mass of about four million solar masses. We know that some of these black holes were already in place very shortly after the big bang, powering the quasars that we see at high redshift, but we have no real understanding yet of how these black holes were formed.
- Myriad galaxies continue to produce myriad young blue stars. Simple calculations tell us that they should have run out of gas by now. How are they replenishing their gas supply?

The public perception may be that science has solved it all, but it has not. We commend an essay written by astronomer Michael Disney:

> Don't be impressed by our complex machines or our arcane mathematics. They have been used to build plausible cosmic stories before—which we had to discard afterwards in the face of improving evidence. The likelihood must be that such revisions will have to occur again and again and again.[3]

We fully agree with Disney that science progresses step-by-step, and in that spirit, we hope that many scientific answers will in the fullness of time be forthcoming.

Disney makes an important (but perhaps too rarely appreciated) point: for most of its existence, our universe has been *opaque*, masked by itself. In physical cosmology, the Planck epoch is the earliest period of time in the history of the universe, from zero to approximately 10^{-43} seconds ($10^{-2} = 0.01$, $10^{-3} = 0.001$, etc.). In this very short, hot, and dense period of the universe, the current laws of physics break down. Given our estimate of the age of the universe to be roughly 10^{17} seconds, some sixty decades or sixty factors of ten in time [$10^{(43+17)} = 10^{60}$] have elapsed since the Planck era. Initially, then, our universe was flooded with radiation. As the universe expanded, the radiation cooled, and about 300,000 years or 10^{13} seconds after the Planck era, the universe became transparent, and galaxies began to form. In logarithmic terms,

---

3. Michael J. Disney, "The Case against Cosmology," *General Relativity and Gravitation* 32, no. 6 (2000): 1134, https://ned.ipac.caltech.edu/level5/Disney/paper.pdf.

galaxies have existed for only four decades in time (from 10^{13} to 10^{17} seconds). Disney poses an insightful thought: How *secure* is our scientific knowledge of the unobservable opaque universe in those first fifty-six out of sixty decades of time? We may get some clues from the Large Hadron Collider, but as Disney says, those early epochs may be "lost too far back in the logarithmic mists of time."[4] *The point Disney emphasizes is that in terms of the history of our universe, current astronomical observations of galaxies span only four out of sixty decades in time!* What a daunting task, Disney therefore argues, to mathematically model the entire history (past and future) of our universe, given that the only astronomical data we have come from a very late fragment in time spanning from 10^{13} to 10^{17} seconds. Disney's essay is one of caution. To put this in perspective, during those mere four decades of time, everything that astronomers have ever observed was produced: all the galaxies, stars, and our solar system were formed, and modern man emerged.

Our response to Peter Atkins would be caution. Atkins may be right, that science has the power to solve all these physical questions, although it still has a long way to go in surmounting the intellectual barriers that we face. Our major issue here, however, is with Atkins's final point: we are concerned not so much with the *power* of science (even if it is sometimes exaggerated) but with the *limits* of science. We interpret Atkins as claiming that science can deal with *any* aspect of our existence, including our existence as spiritual human beings. We would argue otherwise: science is simply not competent to unfold the human story from its spiritual dimension. Let us recall Benjamin Franklin's advice to the chess player, regarding "circumspection, which surveys the whole Chess-board, or scene of action." We believe there is more to the chessboard than the physical pieces. Some scientists are, for whatever reason, blind to the spiritual dimension—it is by grace that our eyes are opened to the spiritual world.

In the words of Scot Bontrager, "Our natural abilities to discern truth about the world ceases with things invisible—*lacking senses to perceive the invisible world* there is no way for us to know truths that lead to our eternal beatitude—the perfection for which we were created."[5]

---

4. Disney, "Case against Cosmology," 1127.
5. Scot C. Bontrager, "Nature and Grace in the First Question of the *Summa*," *Scot Bontrager* (blog), February 1, 2010, https://www.indievisible.org/Papers/Aquinas%20-%20Nature%20and%20Grace.pdf; italics added.

Augustine was very careful never to transgress the limits or boundaries of the book of nature and the book of Scripture. Our viewpoint now, four hundred years after Galileo, is that it is impossible for one book to contradict the other, since both are underpinned by truth. If they appear to disagree, then the problem lies either in the interpretation of scriptural verses, as pointed out by Augustine and emphasized by Galileo in the section above, or in incomplete knowledge about the physical phenomena we observe.

Galileo continued to draw on Augustine:

> *This same saint shows how much the truly wise and prudent Fathers are offended by those who try to uphold propositions which they do not understand by citing passages of Scripture, compounding their original error by producing other passages which they understand even less than the first; he writes:*
>
>> *Rash and presumptuous men bring untold trouble and sorrow on their wiser brethren when they are caught in one of their false and unfounded opinions and are taken to task by those who are not bound by the authority of our sacred books. For then, to defend their utterly reckless and obviously untrue statements, they call upon Holy Scripture, and even recite from memory passages which they think support their position, although they understand neither what they mean nor to what they properly apply.*
>
> *This seems to me to describe exactly those who keep citing passages of Scripture because they are unable or unwilling to understand the proofs and experiments which the author of this doctrine and his followers advance in its support. They do not realize that the more passages they cite and the more they insist that their meaning is perfectly clear and cannot possibly admit any other interpretation than theirs, the more they would undermine the dignity of Scripture (if, that is, their opinions carried any weight) if the truth were then clearly shown to contradict what they say, causing confusion at least among those who are separated from the Church and whom the Church, like a devoted mother, longs to bring back to her bosom. So your Highness can see how flawed*

> is the procedure of those who, in debating questions of natural science, give priority in support of their arguments to passages of Scripture—and often passages which they have misunderstood.
>
> But if they [the opponents of Galileo] really believe and are quite certain that they possess the true meaning of a particular text of Scripture, they must necessarily be convinced that they hold in their hand the absolute truth of the scientific conclusion which they intend to debate, and so must know that they have a great advantage over their opponent who has to defend what is false. The one who is defending the truth will be able to draw on numerous sensory experiences and necessary demonstrations to support their position, while their opponent has to fall back on deceptive appearances, illogical reasoning, and fallacies. . . . But I do not think that their resorting to Scripture to cover up their inability to understand, let alone to answer, the arguments against them, will do them any good.

Galileo continued to rail against his rash and presumptuous opponents who used biblical texts to support their arguments against his scientific ideas. His letter is rather repetitive on this theme, but when we consider what he was facing, with so much at stake, one can surely forgive him for this!

The dispute between Galileo and his critics was presented in terms of the integrity of the Bible, although the primary agenda of the church was to maintain its power. The belief was that the Bible provided a complete account of all that needed to be known. But it is not a scientific textbook, and God is never the author of confusion. Scientific knowledge at the time was very far from complete (and still is in our time), and Galileo once again pointed out the danger of prejudicing people against the dignity of the book of Scripture by using it inappropriately.

Let us again recall the words cited earlier by the astronomer Simon Newcomb, that "we are probably nearing the limit of all we can know about astronomy."[6] *He truly did not know what he did not know.* Our knowledge of the universe remains desperately incomplete. Such is the road of science. As scientists work away on the unknowns, the

---

6. See, for example, Prasenjit Saha and Paul A. Taylor, *The Astronomer's Magic Envelope* (New York: Oxford University Press, 2018), 4.

frontiers of ignorance retreat slowly, year by year. We caution again against invoking the God of the gaps, in which we attribute to God those things that we do not understand, simply because our knowledge of the book of nature is incomplete.

There is a mood in our time that God is at most a retreating God, retreating into the dark corners as science advances. This comes about because of the historical practice of pointing to the God of the gaps. This approach seems to us to be a grievous error. God cannot be known by reason and experiment; he reveals himself through grace to those who seek him. He is not in a competition with science.

Galileo spoke of those "separated from the Church and whom the Church, like a devoted mother, longs to bring back to her bosom." Augustine longed for those outside the grace of God to be brought underneath his mantle of grace. We recall the parable of the prodigal son and the desire of his father to see him return home.

For theologians of the day to think that they were holding, in Galileo's words, "in their hand the absolute truth of the *scientific* conclusion which they intend to debate" (italics added) reminds us of the words of Nobel laureate Rabindranath Tagore: "'The learned say that your lights will one day be no more,' said the firefly to the stars. The stars made no answer."[7]

Galileo stood his ground. In the view of Galileo, his opponents resorted "to Scripture to cover up their inability to understand, let alone to answer, the arguments against them." The converse is also true regarding scientists who attempt to use their reason and experiment in response to scriptural revelation. Theology and science each has its own tools.

Insight must prevail on both sides. The skills to "demonstrate falsity" in present-day science, as in Galileo's time, are limited to an inner circle of cognoscenti. On the one hand, the church needs people who can talk the language of science. We recall that the Vatican does have an observatory, which remains active in research. It is disappointing that much of today's fundamentalist church has so few people who can talk the language of science. But equally, scientists need to be aware that the book of nature has its boundaries; it transgresses the limits of science for scientists to invoke the book of nature in order to assert that God is a delusion.

---

7. Rabindranath Tagore, *Stray Birds* (New York: Macmillan, 1916), 49.

We see before us two cathedrals on the cosmic horizon. They are far apart. Looming in grandeur is the cathedral of science, wherein giants like Galileo, Newton, Pascal, and Einstein have practiced. The foundation of the cathedral of science is the book of nature. The cathedral of God is founded on the book of Scripture. In the mind-set of Galileo at the time of John Milton's visit to him, the books of Scripture and nature were in a harmonious balance. The book of Scripture, according to Milton, had come out of the dustiest of corners, bringing forth radiant truth and light.

If We Seek, We Shall Find

Galileo raised a key issue: that of personal agendas and of personal interest. Many readers might think that today personal interest is out of the equation for scientific matters, but nothing could be further from the truth! For example, professional astronomers cannot use major facilities at an observatory to study any celestial object that they wish; they do not simply wander up to a giant telescope at night to conduct a couple of observations. That requires months of preparation. Time on large telescopes is highly competitive. The granting of time on any gargantuan eye of the sky is the responsibility of a Time Allocation Committee, or TAC. And it is in those committees that personal interests can enter in, particularly if one is requesting time to study an object that a member of the TAC may be studying as well. Astronomy can be *very* territorial! At the time of Galileo, it was theologians with their personal interests and agendas at central focus.

The point to make here is that personalities and agendas can determine scientific outcomes, and the process is not always objective. In the life of Galileo, some men of the cloth flashed their glory; astronomy, too, is filled with some examples of those seeking to be first and, in the process, to receive glory for their discoveries. Trampling over one's peers only for personal reasons would be a crime. In this context, we are reminded of a small extract of a poem by Robert Frost:

Of all crimes the worst
Is the theft of glory.
Even more accursed
Than to rob the grave.[8]

---

8. The lines are an extract from the poem Robert Frost, "Kitty Hawk," *Atlantic Monthly*, November 1957, 52–56. A useful online link to this poem is Richard Stimson, "Robert Frost Writes

The church of Galileo's day had an agenda, to preserve the power of the church at all costs. Their power was all encompassing, even in questions of science. Power plays and the personal interests of Galileo's opponents prevailed.

Galileo recognized that scientific truth exists external to the observer and that it is not in the observer's power to make things true or false. The book of nature transcends human power: individuals have no influence over satellites orbiting Jupiter or over the spots on the seething surface of our sun. Galileo said, "It is beyond the power of any created being to make [propositions] true or false, in defiance of what they are de facto by their own nature." Although there are many questions in astronomy, the wondrous quality of the open book of nature is its transparency. It is beyond imagination that a human being can "instruct" God how to create his universe. As Tagore put it earlier, "'The learned say that your lights will one day be no more,' said the firefly to the stars. The stars made no answer."[9] Most questions have answers, as was already evident to Galileo. There are truths out there to be found, and if we seek, we shall find.

This leads us to the mood of Galileo's time. For Catholics, the church had absolute authority and control over the interpretation of and access to the book of Scripture, but Galileo's point is that no person has power over the book of nature. And this we would argue is the key reason why Pope Urban VIII and others condemned Galileo with such vehemence. They could see the *power* struggle at work here between the church of the time, which told people what to believe, and independent thinkers like Galileo.

It is important to note how Fyodor Dostoevsky portrayed control of theological belief as central to the church at the time of the Spanish Inquisition. Ivan, one of "the brothers Karamazov" speaks of this in the following words: "It's simple lust of power, of filthy earthly gain, of domination—something like a universal serfdom with them as masters—that's all they stand for."[10] But the book of nature is beyond the

---

about the Wright Brothers," *The Wright Stories* (blog), accessed October 2, 2018, http://wrightstories.com/robert-frost-writes-about-the-wright-brothers/. See also Joan St. C. Crane, "Robert Frost's 'Kitty Hawk,'" *Studies in Bibliography* 30 (1977): 241–49.

9. Tagore, *Stray Birds*, 49.

10. Here we are using the translation of *The Grand Inquisitor* by Constance Garnett, Philosophical Investigations, November 13, 2012, http://peped.org/philosophicalinvestigations

church's control. People such as Galileo could decide what to believe for themselves, that the Bible was true, just as what they observed was also true.

The crime of Galileo was the start of a new era. It was people other than theologians—the scientists—who were beginning to set themselves up as alternative authorities in their areas of expertise. The book of nature disarms one of power and of control, such a crucial concept in Dostoevsky's mind. For the theologians, that would have required some humility.

Centuries have passed since the death of Galileo. Today Dostoevsky's grand inquisitor may not be acting in the name of a church but may indeed be parading in the name of the cardinals of science. Only the heresies have changed. What would the grand inquisitor of science have to say to the Creator of the universe? Would he, too, say, "Remove thyself. *Dixi*"? Those who claim to see the complete human story from within the hallways of the cathedral of science alone are spiritually blind to his revelation.

In contrast, we believe that God has visited us in person (John 1). G. K. Chesterton writes,

> But that the Creator was present at scenes a little subsequent to the supper-parties of Horace, and talked with tax-collectors and government officials in the detailed daily life of the Roman empire, . . . that is something utterly unlike anything else in Nature.[11]

The apostle Paul expounds as follows: "The God who said, 'Out of darkness the light shall shine!' is the same God who made his light shine in our hearts, to bring us the knowledge of God's glory shining in the face of Christ" (2 Cor. 4:6 GNB). God of the macrocosm is God of the microcosm, shining into our hearts, if invited in. We can never *know* God by mere scientific analysis but only by grace. By grace, the spiritually blind on earth are given sight to behold him, God incarnate.

/dostoyevskygrandinquisitor/. See also F. M. Dostoevsky, *The Grand Inquisitor*, trans. S. S. Koteliansky (London: Elkin Mathews & Marrot, 1930), 30.

11. G. K. Chesterton, *The Everlasting Man* (London: Hodder and Stoughton, 1925), 266.

In the words of the great cosmologist Allan Sandage (see fig. 13),

> Science is not the way to find *truth*, but only probable truth. You get only an approximation that always changes, and there are no absolutes. You don't read a physics textbook written 300 years ago to learn about physics, and I expect that 100 years from now you're not going to read the *Astrophysical Journal*[12] [of today] to learn about astronomy. Science is the only self-correcting human institution, but it is also a process that progresses only by showing itself to be wrong.[13]

Furthermore, as noted by Sandage, because of the limitations of its method by reason alone, it is beyond the scope of science to explain and understand *everything* about reality. God is not subject to Einstein's field equations, to Lie algebras, or to Erwin Schrödinger's wave equation (all concepts in cosmology and theoretical physics).

It is the mystery of Chesterton's poem "Gloria in Profundis" (Glory to God in the lowest):

> Who is proud when the heavens are humble,
> Who mounts if the mountains fall,
> If the fixed stars topple and tumble
> And a deluge of love drowns all—
> Who rears up his head for a crown,
> Who holds up his will for a warrant,
> Who strives with the starry torrent,
> When all that is good goes down?

Jesus is *the* Starry Messenger. For him all stars have shone.[14]

The Hinges of the Earth

Galileo concluded his letter to the Duchess by drawing on Proverbs 8:26:

> *O God, whose hand hath spread the sky,*
> *and all its shining hosts on high,*

---

12. *Astrophysical Journal* is one of the world's premier peer-reviewed research journals devoted to recent developments, discoveries, and theories in astronomy and astrophysics.
13. Quoted in Alan Lightman and Roberta Brawer, *Origins: The Lives and Worlds of Modern Cosmologists* (Cambridge, MA: Harvard University Press, 1992), 82. See also Allan Sandage, "A Scientist Reflects on Religious Belief," LeadershipU, accessed October 2, 2018, http://www.leaderu.com/truth/1truth15.html.
14. G. K. Chesterton, "Gloria in Profundis" (London: Faber & Gwyer, 1927). Public domain.

*and painting it with fiery light,
made it so beauteous and so bright:*

*Thou, when the fourth day was begun,
didst frame the circle of the sun,
and set the moon for ordered change,
and planets for their wider range.*

They could also say that the word "firmament" is literally correct for the starry sphere and for everything which is beyond the revolutions of the planets, for in the Copernican system this is totally firm and immobile. And since the Earth moves in a circle, when they read the verse, "Before he had made the Earth and the rivers, and the hinges of the earth," they might think of its poles, for it seems pointless to attribute hinges to the terrestrial globe if it does not turn on its axis.

The Douay-Rheims Bible translates Proverbs 8:26 as follows: "He had not yet made the earth, nor the rivers, nor the poles of the world." Instead of the word "poles," the Latin words *cardines orbis terrae* may be translated "the hinges of the terrestrial orb."[15]

In 2009 we both saw an excellent play in Boston, *The Hinge of the World*, written by Richard Goodwin.[16] The play focuses on the historic battle between Galileo and Pope Urban VIII. Goodwin terms this the "battle for the soul of the world." As depicted so dramatically, Galileo's struggle was not a "long walk to freedom" (to borrow the title of Nelson Mandela's famous book) but rather a "long walk to house arrest." The shackles would only be fully broken in the fullness of time, when the common person could finally read banned scientific books.

While the struggle of Galileo is often discussed from the point of view of suppressed science and church control, we were impressed by Pope Urban VIII's foresight and his predicament as custodian of the Roman church in the play. He said, "You do not deny God, Galileo.

---

15. For an online Latin Vulgate rendition of Proverbs 8, see Vulgate, Christian Classics Ethereal Library, accessed October 10, 2018, http://www.ccel.org/ccel/bible/vul.Prov.8.html
16. Richard N. Goodwin, *The Hinge of the World: A Drama* (New York: Farrar, Straus and Giroux, 1998).

We are not threatened by denial. We are accustomed to denial. It is worse. You would make God unnecessary."[17]

The interpretation of the Bible and who could read it were under the watchful eye of the Inquisition. It was not a private matter but under the church's rigid control. The Venetian Interdict issued by the pope against the Republic of Venice in 1606 tightened the grip. What was the accessibility of the entire New Testament in Italian to ordinary citizens in the Republic of Venice in the period before Galileo presented his spyglass to the Venetian Senate in 1609? Author Arabella Georgina Campbell provides rich insight. She refers to a sermon delivered by the friar Fulgenzio Micanzio in this time: "If the Saviour were now to ask the question 'Have ye not read?' all the answer you could make would be 'No, we are forbidden to do so.'" Campbell continues:

> And it is a fact equally well vouched for, that when the Church resounded with the same demand that Pontius Pilate the Roman Governor made to the Holy Saviour, "*What is truth?*" Fulgenzio took a New Testament from his pocket and told his [audience] that after long search he had found it there. He held the precious volume up with outstretched hands in the sight of all, "*But,*" he exclaimed as he returned the book to its place, "*the book is prohibited!*"[18]

Campbell references the first translation of the Bible *into Italian* from the Hebrew and Greek by Giovanni Diodati (1576–1649). It appeared in 1603 (and was expanded with notes in 1607), only seven years before the publication of Galileo's *Sidereus Nuncius* in Venice in 1610.

We see the world swinging on two hinges. The first hinge belongs to the book of Scripture, which, to paraphrase Milton, is understandable not only to the wise and learned but also to the simple, the poor, the babes. The first hinge requires spiritual revelation and divine grace. The second hinge is the book of nature: God's created world. It is on both these hinges that we see the fullness of human existence. May we

---

17. Goodwin, *Hinge of the World*, 196.
18. Arabella Georgina Campbell, *The Life of Fra Paolo Sarpi* (London: Molini and Green, 1869), 174.

never place the first hinge on our personal index of forbidden books and thus miss out on the clearest revelation of God's love.

Our commentary on Galileo's *Letter to the Grand Duchess Christina of Tuscany* nears its end. At the time of Galileo's death in 1642, the Grand Duke of Tuscany wished to have him buried in the Basilica of Santa Croce in Florence, but Pope Urban VIII objected on the grounds that the church had denounced Galileo as a suspected heretic. Nevertheless, his legacy lives on. Why? He stood for truth. Crushed truth will be resurrected. The humiliation of Galileo, clad in his shirt of penitence, kneeling in the presence of his assembled judges in Rome in 1633, and his subsequent life sentence of house arrest by the Inquisition cemented the beginning of modern science.[19] No amount of terrestrial darkness—no earthly storm—could suppress the light from the luminous starry vaults of the *Via Lactea*, or Milky Way. The stamp of the Inquisition gave value to the "coin of life" of Galileo, to paraphrase Rabindranath Tagore.[20]

His struggle is our struggle, to seek that which is truly precious and to restore the balance between the two hinges. However violent the storm, light from the book of Scripture beams forth. As Jesus said, "Again, the kingdom of heaven is like treasure hidden in a field, which a man found and hid; and for joy over it he goes and sells all that he has and buys that field" (Matt. 13:44 NKJV). Only God can remove the shroud or film of ignorance that may cover our eyes from seeing that first hinge—his kingdom—a treasure of inestimable price, lying hidden in a field. Let us, like Kepler, Galileo, and Copernicus, be fully conscious of our role as scientists, never transgressing the boundaries of the book of nature and fully embracing the kingdom of heaven.

The hinge on which the kingdom of heaven turns is the incarnation. We cannot think outside time, so God enters our confines of time. We cannot think outside space, so God enters our world of space. In the course of human history, the Logos ("I am") entered our confines of space and time: "And the Word was made flesh, and dwelt among

---

19. For the full text of Galileo's sentence, see Giorgio de Santillana, *The Crime of Galileo* (Chicago: University of Chicago Press, 1955), 306–10. Also available online at Douglas O. Linder, "Papal Condemnation (Sentence) of Galileo," Famous Trials, accessed September 13, 2018, http://www.famous-trials.com/galileotrial/1012-condemnation.

20. Tagore, *Stray Birds*, 31.

us, (and we beheld his glory, the glory as of the only begotten of the Father,) full of grace and truth" (John 1:14 KJV).

We give G. K. Chesterton the final word of this section:

> Right in the middle of historic times, there did walk into the world this original invisible being; about whom the thinkers make theories and the mythologists hand down myths; the Man Who Made the World.[21]

---

21. Chesterton, *Everlasting Man*, 266.

PART 2

HISTORICAL VIGNETTES

9

A Moon of Glass from Murano, Venice

Our first historical vignette results from a question that was posited to us by our colleague Bruce Elmegreen in New York: "Do any glass factories in Murano (Venice) still exist today whose family history dates back to the time of Galileo or even earlier?" The answer is contained in this story.

The first known telescope (or spyglass) was invented in 1608 in the Netherlands, and the news spread quickly throughout Europe. By late March 1609 telescopes had arrived at the court of Albert, Governor of the Hapsburg Netherlands, where they were reproduced and sent to Emperor Rudolf II in Prague and to the king of Spain in Madrid. In May 1609 a certain "Frenchman" offered the Count of Fuentes in Milan a specimen of the spyglass. Thomas Harriot constructed a spyglass in England in July 1609 and used it to look at the moon.[1]

In July 1609 our story shifts from the invention of the spyglass in Holland to the Republic of Venice. A key figure in this interlude is the Venetian Paolo Sarpi. In their book *Galileo's Telescope*, authors Massimo Bucciantini, Michele Camerota, and Franco Giudice introduce Sarpi as follows:

> *This* is the Sarpi who intrigues us: the man who, before becoming *totus historicus* and *politicus*—director and prime actor in

---

1. Stephen Pumfrey, "Harriot's Maps of the Moon: New Interpretations," *Notes and Records of the Royal Society* 63, no. 2 (2009): 163–68.

the battle against the Papal Interdict (1606–1607) and then the famous condemned author of the *Istoria del Concilio Tridentino* (1619)—tirelessly probed scientific matters and who, in the story of the spyglass, was destined to play a key role that has often been underestimated in studies on the history of the telescope and in some cases overlooked entirely.[2]

The Republic of Venice was in conflict with the pope, and Sarpi radically argued against papal authority in matters of state. Sarpi had received news of the telescope already in November 1608. At that time, it was rumor only; he did not have a specimen before him to examine the secret of the spyglass. The next year, this was no longer a rumor in Venice. On July 21, 1609, Sarpi penned these words: "A spyglass has arrived that makes far away things visible. I greatly admire it because of the beauty of the invention and its skillful craftsmanship, but I find it worthless for military purposes, either on land or at sea."[3] Paula Findlen elaborates: "Sarpi used his extensive political network to collect news of the spyglass; he was frequently seen with Galileo, discussing technical problems, identifying the best artisans and materials, and observing the heavens."[4]

In a short time, the secret was open:

> Thanks to the work of master craftsmen who instantly grasped that there was money to be made, the *occhiali in canna* or *trombette* became an open secret and could be purchased in many cities, proving not only their widespread circulation but also the poor quality of many specimens.[5]

Therein lay the key. The lenses were of poor quality, but the instrument was a great novelty.

Sarpi was the scientific interlocutor of Galileo.[6] But time was of the essence for Galileo to construct spyglasses of his own. We read:

---

2. Massimo Bucciantini, Michele Camerota, and Franco Giudice, *Galileo's Telescope: A European Story*, trans. Catherine Bolton (Cambridge, MA: Harvard University Press, 2015), 34.
3. Bucciantini, Camerota, and Giudice, *Galileo's Telescope*, 35.
4. Paula Findlen, "The Spyglass and the Astronomer: Seeing Galileo in Perspective," *Los Angeles Review of Books*, October 3, 2015, https://lareviewofbooks.org/article/the-spyglass-and-the-astronomer-seeing-galileo-in-perspective/.
5. Bucciantini, Camerota, and Giudice, *Galileo's Telescope*, 37.
6. Mario Biagioli, "Did Galileo Copy the Telescope? A 'New' Letter by Paolo Sarpi," in *The Origins of the Telescope*, ed. Albert van Helden, Sven Dupré, Rob van Gent, and Huib Zuidervaart (Amsterdam: Knaw, 2010), 221–22.

The week of September 22 to 29 [1609] proved fateful for Galileo. It would mark the beginning not only of a new trade, that of a lens maker admired and sought after throughout Europe, but also of a new life. At forty-six years of age, he was about to see his existence change dramatically. The *perfection* of the Dutch spyglass and its repurposing into an astronomical instrument would completely transform his [Galileo's] routine and work. . . . People in town spoke of nothing else.[7]

Galileo presented a spyglass to the Senate of Venice on August 24, 1609; its tube was approximately 60 cm long and 42 mm in diameter, with a magnification power of 8. Everyone was "abuzz with talk" about Galileo's being able to greatly surpass in quality the spyglass that had been developed in Holland. We do not know precisely when Galileo commenced work on his first spyglass. But we do know not only that Sarpi was well informed about the spyglass but also that he had been actively involved in its construction.

A landmark in the birth of modern astronomy occurred on March 13, 1610: the publication in Venice of Galileo's legendary *Sidereus Nuncius* (or *Sidereal Messenger*).[8] It was the first published scientific work based on observations that Galileo had made through his telescopes, and the treatise contained Galileo's early observations of a mountainous moon, myriad stars in our Milky Way galaxy that were beyond the grasp of the naked eye, and four Medicean moons that orbited the planet Jupiter.

On March 16, 1610, only three days after the publication of the *Sidereus Nuncius*, Sarpi wrote a letter to Jacques Leschassier, which alludes to spyglasses that Sarpi had made. That letter, in part, reads,

> As you know, this instrument is composed of two lenses (which you call *lunetes*), both of which spherical, one with a convex surface and the other concave. . . . We [Sarpi and his entourage] made one from a sphere with a diameter of six *piedi*, and the one from another sphere a digit smaller in diameter.[9]

---

7. Bucciantini, Camerota, and Giudice, *Galileo's Telescope*, 38; italics added.
8. Galileo Galilei, *Sidereus Nuncius*, or *The Sidereal Messenger*, trans. Albert van Helden (Chicago: University of Chicago Press, 1989).
9. Quoted in Bucciantini, Camerota, and Giudice, *Galileo's Telescope*, 43.

And now, a note of strategy and intrigue. When Galileo wrote up his section about the history of the spyglasses used for his observations in the *Sidereus Nuncius*, all credit belonged to Galileo himself. Sarpi was excluded from any mention.

Galileo was fiercely territorial about his role in perfecting the telescope, and he did not wish to acknowledge the names of others.[10] Creating the perception that Galileo himself heroically developed such instruments in total isolation was clearly the goal, but it is false. Valuable insight is provided here by Bucciantini, Camerota, and Giudice. The omissions are glaring, and they are careful to note this: "The *Sidereus* is also a work in which intentional gaps and voids stand out. ... When we read it with disenchanted eyes and, so to speak, *ex parte veneta*, what emerges is an equally systematic and conscious attempt to conceal people, words, and events."[11]

Here we find Sarpi, who had "shared nearly twenty years of endless conversations about nature and humankind" with Galileo, betrayed by Galileo.[12] Was Sarpi not Galileo's closest scientific interlocutor while Galileo was in Padua? What a dramatic turn of events, as expressed by Bucciantini, Camerota, and Giudice: "There is no doubt that the collaboration of Sarpi and his entourage was decisive, *although Galileo never publicly acknowledged this.*"[13]

In summary, the crucial role played by the Venetian Paolo Sarpi

> is amply demonstrated . . . by the fact that he conducted his celestial observations at the monastery of Santa Maria dei Servi. . . . The fact remains that the history of the spyglass does not travel along a single vector or one-way path but, as Venice demonstrates, is an archipelago of overlapping human and intellectual events: a collective story that must be recounted in its simultaneity in order to be reconstructed and reassembled piece by piece.[14]

---

10. Such practices are still not uncommon in some scientific circles today. Who actually makes the discovery and who subsequently gets the credit is a subject of great interest in the sociology of science.
11. Bucciantini, Camerota, and Giudice, *Galileo's Telescope*, 47.
12. Bucciantini, Camerota, and Giudice, *Galileo's Telescope*, 195.
13. Bucciantini, Camerota, and Giudice, *Galileo's Telescope*, 36; italics added.
14. Bucciantini, Camerota, and Giudice, *Galileo's Telescope*, 44, 52.

Which materials did Galileo use to manufacture his lenses? The focal point in our historical narrative is once again Venice.

We were privileged to be guest researchers at the Biblioteca Nazionale Centrale di Firenze (Florence). It is the largest public national library in Italy and one of the most important in Europe. The library houses multitudes of handwritten documents by Galileo. Among the archives is a well-known shopping list (1609) in the handwriting of Galileo for a forthcoming trip to Venice. It includes

> two artillery balls (*palle d'artiglieria*), a tin organ pipe (*canna d'organo di stagno*), polished German glass (*vetri todeschi spianati*), rock crystal (*cristallo di monte*), pieces of mirror (*pezzi di specchio*), iron bowls (*scodelle di ferro*), Tripoli powder (*tripolo*), an iron plane (*ferro da spianare*), Greek pitch (*pece greca*), felt, mirror to rub (*feltro, specchio per fregare*) and fulled wool (*follo*).[15]

Anyone with a knowledge of the grinding of lenses would recognize these as key items on Galileo's shopping list for the construction of a telescope.[16] Galileo had to perfect the lenses himself.

Bucciantini, Camerota, and Giudice offer us detailed insights into the processes involved:

> Finding the materials [Galileo] needed was merely the first step. . . . The real work began when a disk was cut from the selected sheet of glass and an iron plane was used like a file to remove the excess along its edges. At this point, to obtain a lens in the desired shape, the glass disk had to be ground on a suitable mold: artillery balls for concave lenses and iron bowls . . . for convex ones.[17]

Abrasive Tripoli powder was used to create a smooth surface, which became transparent with the application of yet finer Tripoli powder. We furthermore read that it was "polished until it shone, using a piece of felt or fulled wool and going to great lengths to avoid

---

15. Bucciantini, Camerota, and Giudice, *Galileo's Telescope*, 59.
16. Toward the end of November 1609, Galileo had established his own optical workshop in Padua.
17. Bucciantini, Camerota, and Giudice, *Galileo's Telescope*, 62.

altering its curvature."[18] And of course, organ pipes themselves could be used as tubes for telescopes.

In order to minimize defects in the lenses themselves, Galileo used only the inner parts of the lenses: this reduced the effect of bubbles and inclusions in the glass and of imperfections in the grinding of the lenses. His lenses were ground and polished oversize to better control the actual shape of the surface, and Galileo used opaque diaphragms to let through only the light from the inner parts of the lens that gave the best images.

As gleaned from Galileo's shopping list, Venice was an ideal city to secure both the raw materials, such as glass, and the necessary grinding tools for his lenses. Venice is, of course, world renowned for the manufacture of glass, and we now return to the question raised by Bruce Elmegreen, whether any glass factories in Venice still exist today whose family history dates back to the time of Galileo, or possibly earlier.

How fascinating that glass should feature on Galileo's list of items to buy in Venice. A boat ride from Venice to Murano (a series of islands separated by canals but connected to one another by bridges, slightly less than one mile north of Venice) is a mere ten minutes or so from the city of Venice itself. In her book with photographer Norbert Heyl, the historian Rosa Barovier Mentasti elaborates:

> The peculiar feature that distinguishes it from the larger city is that behind the façades of the picturesque old buildings lined up along the canals, the fires continue to burn in the furnaces in which the works of art of the Venetian glass-making tradition have been shaped for at least eight centuries.[19]

In visiting Murano, one quickly senses that the artistry of Murano glass is a multigenerational world of discovery, innovation, and networking that have all fostered the industry.

One of the most famous glass factories in Murano today is known as Barovier and Toso. The family name of Barovier is renowned for the famed *cristallo* (a completely transparent, colorless glass, without the slight yellow or greenish color originating from iron oxide impurities),

---
18. Bucciantini, Camerota, and Giudice, *Galileo's Telescope*, 62.
19. Norbert Heyl and Rosa Barovier Mentasti, *Murano: The Glass-Making Island*, trans. Clare Loraine Walford (Ponzano Veneto: Vianello Libri, 2006), 15.

which comes from these islands in the Venetian lagoon. The invention of *cristallo* glass is attributed to Angelo Barovier around 1450.

Meeting Rosa Barovier in Venice recently was a remarkable experience; she is undoubtedly one of the most important historians of Venetian glass today. The history of her extended family and of the manufacture of glass in Venice dates back, long before the time of Galileo, to the fourteenth century. Apart from being a Barovier, born and raised in Murano, she has the most profound insight into the world of glass makers at the time of Galileo, being also intimately acquainted with the writings of Antonio Neri (1576–1614) of Florence.[20]

During our interview, Rosa Barovier painted a fascinating picture of a fifteenth-century glassmaker who stole the recipes of glassmaking from Marietta Barovier, daughter of the famed Angelo Barovier. His name was Giorgio Ballarin, a crippled and poor young man at the time he worked for the Barovier family. He looked so naïve, and Marietta Barovier trusted him. She prepared glassmaking recipes in front of him. Little could Marietta Barovier have guessed that Giorgio Ballarin would give the recipes to Angelo Barovier's rival, who subsequently became one of the leading glassmakers of Murano.

Rosa Barovier also recalled the name of Bortolo d'Alvise, who ran a glass furnace in Murano at the sign of the "Tre Mori," or "Three Moors." The Grand Duke of Florence from 1537 to 1574, Cosimo I, was determined to bring the art of Venetian glass to that city. How could glass trade secrets from Murano be brought to Florence? He secured the services of Bortolo d'Alvise, who, with two colleagues, moved from Murano to Florence in 1569. Of course, d'Alvise took all his trade secrets with him.

Intellectual property was indeed a problem of vast proportions already centuries ago. Joanna Kostylo emphasizes this point:

> The possibility of the flight of artisans with the consequent diffusion of their techniques was a problem that had haunted the Venetian government for centuries, despite increasingly severe penalties

---

20. Neri's book *L'Arte Vetraria*, published in 1612, is possibly the most famous early book on the manufacture of different kinds of glass. For an English translation, see Antonio Neri, *The Art of Glass*, trans. Christopher Merrett, ed. Michael Cable (Sheffield, UK: Society of Glass Technology, 2006).

for artisans caught jeopardising the Venetian monopoly. . . . Such extreme remedies, however, were not unique to Venice. In Genoa, the city most injured by artisan emigration, the authorities offered in 1529 a reward up to two hundred ducats to anyone who killed a fugitive artisan.[21]

And now, some fascinating insights by Rosa Barovier about Galileo. She reminded us of a link Galileo had established between the moon and glass. Rosa read to us a section of the text of the *Sidereus Nuncius* in which Galileo compared the appearance of the moon through his spyglasses to that of cracked glass, or "ice glass" (Ital. *vetro a ghiaccio*; Lat. *Glaciales Cyathi*). The section is also cited by Eileen Reeves:

> This lunar surface, which is decorated with spots like the dark blue eyes in the tail of a peacock, is rendered similar to those small glass vessels which, plunged into cold water while still warm, crack and acquire a wavy surface, after which they are commonly called *cyathi*, or "ice-cups."
>
> . . . Galileo understood that the vast part of his audience had never used a telescope, and that clear engravings and forceful imagery would be the only means of persuading them that the moon was anything other than a polished sphere. At the same time, the comparisons were meant to be of some aesthetic appeal to such readers: the spots on the lunar surface were as beautiful as those on the peacock's tail, and the unending chain of peaks, valleys, and craters, like that on fine Venetian glassware, not a random production of nature but a pattern imposed by that *Miglior Fabbro*, the Best of Artisans. The fact that Galileo had first learned of the telescope while in the Veneto, that he used Venetian glass for its lenses, and that he had given one of these instruments to the Venetian Senate and had demonstrated its military potential to them, may also have encouraged him to use a comparison that evoked that republic and one of its important exports [glass].[22]

---

21. Joanna Kostylo, "From Gunpowder to Print: The Common Origins of Copyright and Patent," in *Privilege and Property: Essays on the History of Copyright*, ed. Ronan Deazley, Martin Kretschmer, and Lionel Bently (Cambridge: Open Book, 2010), 34.
22. Eileen Reeves, *Painting the Heavens: Art and Science in the Age of Galileo* (Princeton, NJ: Princeton University Press, 1997), 150.

Could we ever find a "miniature glass moon" made of cracked glass on the islands of Murano—the cracked glass that was in the mind of Galileo when he described the appearance of our moon?

Intrigue and much surprise lay in store for us that day. We had reached a glass factory in Murano by boat, not open to the public. To our amazement, Rosa had organized for us to meet a Murano glass master, Andrea Zilio. Next came the magic moment. Before our eyes, a sphere of cracked *vetro a ghiaccio* was produced by Zilio and his assistants, made especially for us (see fig. 14). A miniature moon of glass, which originated from fiery furnaces in Murano, raging at temperatures of approximately 1000 degrees centigrade! From Florence, where we personally examined the original manuscript of the *Sidereus Nuncius*, to the Veneto, to then holding a sphere of cracked glass in our hands—it was an unforgettable experience.

On the isles of Murano is an analogy not only to our moon but also to our galaxy. Before leaving Murano with Rosa, we held the most delicate spiral patterns, all inside glass. Rosa was showing us examples of *reticello* glass (also known as *redeselo* glass), which Galileo would presumably have seen (as *reticello* glass was made in Murano from 1549 onward).

Within a transparent glass base is embedded a network of crisscrossing threads of opaque glass, which form a lattice of diamond-shaped pockets. As Paul Engle comments, "The overall effect is reminiscent of fine lace or of fishing net, both of which are strongly evocative in Venetian culture."[23]

A veritable journey to Murano it was, from the microcosm to the macrocosm—from spirals produced inside *reticello* glass in the glass furnaces in Murano to the multitudes of spiral-shaped galaxies known to astronomers today, including the spiral galaxy observed by Galileo himself: the *Via Lactea*, or Milky Way.

We conclude our story by returning from Murano to the city of Venice itself. The sun was about to set; after bidding farewell to Rosa Barovier, we took a twilight walk along the Strada Nova in Venice. We found ourselves in front of a statue: a statue of Paolo Sarpi himself.

---

23. Paul Engle, "Reticello," *Conciatore* (blog), March 13, 2015, https://www.conciatore.org/2015/03/reticello.html.

It was as if Sarpi was trying to remind us of something. What might the silent messenger have been trying to tell us? Could it be that here stands the man—Paolo Sarpi, theologian, scientist, historian, and statesman—who had not only survived two assassination attempts but who had possibly even shown Galileo a spyglass for the first time in July 1609?[24]

---

24. Biagioli believes that the date of Galileo's visit to Sarpi in Venice may have been about July 19–20, 1609. On July 21, 1609, Sarpi had written that "a spyglass has arrived that makes faraway things visible." See Biagioli, "Did Galileo Copy the Telescope?," 221.

10

A Troubled Dinner in Tuscany

A second intriguing historical vignette retraces the background of Galileo addressing his *Letter to the Grand Duchess Christina of Tuscany*. Was a moving earth (as asserted by Copernicus in 1543) compatible with the book of Scripture? Our discourse takes us to a troubled luncheon in Tuscany, attended by the Grand Duchess of Tuscany as well as by Benedetto Castelli, a Benedictine monk who studied with Galileo from 1604 to 1606. Maurice Finocchiaro sets the scene:

> In December 1613, the Grand Duchess Dowager Christina confronted one of Galileo's friends and followers named Benedetto Castelli, who had succeeded him in the chair of mathematics at the University of Pisa; she presented Castelli with the biblical objection to the motion of the earth. This was done in an informal, gracious, and friendly manner, and clearly as much out of genuine curiosity as out of worry. Castelli answered in a way that satisfied both the duchess and Galileo, when Castelli informed him of the incident.[1]

Galileo responded to Castelli in a famous letter written in 1613 explaining the harmony between the book of nature and the book of Scripture. This letter to Castelli was the foundation for Galileo's *Letter to the Grand Duchess Christina.*

As noted by the Galilean scholar Annibale Fantoli,

---

1. Maurice A. Finocchiaro, trans. and ed., *The Trial of Galileo: Essential Documents* (Indianapolis: Hackett, 2014), 16, 17.

It is difficult to imagine that Galileo would have written this letter exclusively for Castelli. He must, in fact, have seen that it would have had a more general audience in the fond hope that it would help to alleviate the growing opposition and worries about Copernicanism [a moving as opposed to an immobile earth] which were founded on the literal interpretation of certain passages of Scripture. In fact, in a very short time copies of the Letter to Castelli began to circulate. But the outcome was quite the opposite of the hoped for result.[2]

The drama continued to unfold. In December 1614 a Dominican friar named Tommaso Caccini "preached a Sunday sermon against mathematicians in general, and Galileo in particular, on the grounds that their beliefs and practices contradicted Scripture and were thus heretical," writes Finocchiaro.[3]

In February 1615 Galileo's letter to Castelli was transmitted to the Roman Inquisition by another hostile Dominican friar, Niccolò Lorini. It was sent to Cardinal Paolo Emilio Sfondrati, who, as Finocchiaro explains, was "head of the Congregation of the Index and member of the Congregation of the Holy Office in Rome."[4]

These were dangerous times; the watchful eye of the Inquisition was everywhere. At stake was the authority of the church of Rome. Was the geography of the heavens to be determined by literal biblical interpretations enforced by the Inquisition? Galileo argued otherwise, but the opposition was fierce. On February 16, 1615, Galileo penned these words to his friend Monsignor Piero Dini: "They are making an uproar about [the letter to Castelli]; from what I hear, they find many heresies in it and, in short, they have opened a new front to tear me to pieces."[5]

The threat of a moving earth was proving too much for the church. The heat from Galileo's enemies grew in intensity. Copernicanism celebrated a moving earth in a sun-centered, or heliocentric, system

---

2. Annibale Fantoli, *Galileo: For Copernicanism and the Church*, trans. George V. Coyne, Vatican Observatory Publications, Studi Galleiani 3 (Notre Dame, IN: University of Notre Dame Press, 1994), 131.
3. Finocchiaro, *Trial of Galileo*, 17.
4. Finocchiaro, *Trial of Galileo*, 17.
5. Quoted in Fantoli, *Galileo*, 134.

(named after Nicolaus Copernicus, author of the masterpiece *De revolutionibus orbium coelestium*, or *On the Revolutions of the Heavenly Spheres*), and this scientific philosophy was to be officially suppressed by a decree issued by the Congregation of the Index only one year later, in March 1616.

We step back a year. Valuable insight is given by Finocchiaro:

> In March [1615], Caccini made a personal appearance before the Roman Inquisition. In his deposition, he charged Galileo with suspicion of heresy, based not only on the content of the letter to Castelli, but also on the *Sunspots* book (1613). . . .
>
> In the meantime, Galileo was writing for advice and support to many friends and patrons who were either clergymen or had clerical connections. He had no way of knowing about the details of the Inquisition proceedings, which were a well-kept secret; but Caccini's sermon had been public, and also he was able to learn about Lorini's written complaint.
>
> Galileo also wrote and started to circulate privately three long essays on the issues. One, the *Letter to the Grand Duchess Christina*, dealt with the religious objections and was an elaboration of the letter to Castelli, which was thus expanded from eight to forty pages.[6]

Why did his letter to Castelli create the uproar it did? Fantoli explains:

> In the *Letter to Castelli* Galileo had entered into theological matters and had pretended, even though he was a simple scientist, to deal with matters of Biblical interpretation. That was extremely serious (the other fathers also agreed) because it set up an example of the kind of private interpretation of Holy Scripture which the Catholic Church had condemned.[7]

Fantoli writes,

> Without a doubt Galileo was encouraged in his Copernican campaign as well as in the composition of his *Letter to Christina of Lorraine* by the publication of a work by the Carmelite theologian,

---
6. Finocchiaro, *Trial of Galileo*, 17.
7. Fantoli, *Galileo*, 132.

Antonio Foscarini. This publication was entitled: *Letter of the Reverend Father Master Antonio Foscarini, Carmelite, on the opinion of the Pythagoreans and of Copernicus concerning the mobility of the Earth and the stability of the Sun and the new Pythagorean system of the world, etc.* It reproduced a letter sent by Foscarini himself to the Superior General of the Carmelites. On 7 March [1615] Cesi[8] had sent a copy to Galileo with an accompanying letter in which he commented:

> ... a work which certainly could not have appeared at a better time, unless to increase the fury of our adversaries is damaging, which I do not believe.

In his work Foscarini gave importance above all to the inadequacy and unlikelihood of the system of Ptolemy. He then spoke of Galileo's discoveries thanks to which the Copernican hypothesis now appeared to be more acceptable since it was simpler and fit the observations better.[9]

It was critical now for Galileo to lay out his arguments and build up a broad and influential base of support. Galileo sought to protect himself under the shield of the house of Tuscany. The Grand Duchess of Tuscany would be an influential route to her son the Duke, who had appointed Galileo to his post in Pisa and was Galileo's patron. What better person for Galileo to dedicate his letter of 1615 to than the Grand Duchess herself? This is how the *Letter to the Grand Duchess Christina*, which forms the framework of our book, was birthed.

The original *Letter to the Grand Duchess* was never found, but we saw handwritten copies of it in the Biblioteca Nazionale Centrale di Firenze in Florence. In fact, over sixty copies remain. Why were so many handwritten copies made? Handwritten copies spread like a bushfire from one reader to the next. Those who wished to read the letter had to borrow it from friends and transcribe it. Printing such a letter in Italy in 1615—with its high praise for Copernicus—may clearly *not* have been prudent. Galileo's *Letter to the Grand Duchess*

---

8. Federico Cesi was a well-placed colleague of Galileo.
9. Fantoli, *Galileo*, 138–39.

was only published more than *two decades* later (in 1636). Not in Italy, however—it first appeared in print in Strasbourg.

After the writing of his *Letter to the Grand Duchess Christina* in 1615, Galileo continued to weather the raging storms of church control, including the banning of Copernicus's book the very next year, in 1616. In the final analysis, Galileo represented far too much of a threat, and he was summoned, tried, and condemned before the Inquisition. In a letter dated January 15, 1633, Galileo himself reflected back on his writing of the *Letter to the Grand Duchess*:

> Many years ago, at the beginning of the uproar against Copernicus, I wrote a very long essay showing, largely by means of the authority of the Fathers, how great an abuse it is to want to use the Holy Scripture so much when dealing with questions about natural phenomena, and how it would be most advisable to prohibit the involvement of Scripture in such disputes; when I am less troubled, I shall send you a copy. I say less troubled because at the moment I am about to go to Rome, summoned by the Holy Office, which has already suspended my *Dialogue*.[10]

Galileo was referring to his book *Dialogue concerning the Two Chief World Systems*, or *Dialogo sopra i due massimi sistemi del mondo*, which appeared in 1632 and in which he compared the Copernican cosmos to the traditional earth-centered universe of Ptolemy.

Galileo continued: "From reliable sources I hear that the Jesuit Fathers have managed to convince some very important persons that my book is execrable and more harmful to the Holy Church than the writings of Luther and Calvin."[11]

His sentence—that of house arrest—was subsequently signed by seven cardinals. While one of the most famous trials in the history of science had ended, the earth quietly moved in its orbit around the sun.

---

10. Quoted in Maurice A. Finocchiaro, *The Galileo Affair: A Documentary History*, California Studies in the History of Science (Berkeley: University of California Press, 1989), 225.
11. Finocchiaro, *Galileo Affair*, 225.

11

Winning Back Trust

Astronomy and the Vatican

Our third historical vignette focuses on why the Vatican has its own observatories at the pope's summer residence in Italy and in Arizona, given Galileo's sentence in 1633.[1] There is no observatory at the residence of the archbishop of Canterbury, for example.

Astronomy in the Vatican goes back to the time of Galileo's childhood. In 1576 Pope Gregory XIII commissioned the Torre dei Venti, or the Tower of the Winds, in the Vatican. This was known as the Gregorian Observatory, or Gregorian Tower, rising 240 feet high. At the time of construction, there were, of course, no telescopes. What did it thus contain? An exceedingly informative book has been written by the physicist and priest Sabino Maffeo and translated from Italian into English by one of our esteemed astronomy colleagues at the Vatican, George Coyne. Maffeo explains:

> But the most important item in the room [of the calendar] of scientific, and specifically of astronomical, interest is the meridian line[2]

---

1. For the full text of Galileo's sentence, see Giorgio de Santillana, *The Crime of Galileo* (Chicago: University of Chicago Press, 1955), 306–10. Also available online at Douglas O. Linder, "Papal Condemnation (Sentence) of Galileo," Famous Trials, accessed September 13, 2018, http://www.famous-trials.com/galileotrial/1012-condemnation.

2. When our sun crosses the celestial meridian each day, at approximately noon, it is at its greatest altitude (angular distance) above our horizon.

constructed by Danti himself. A hole is located in the south wall at a height of about 5 metres and the artist placed it right in the mouth of a genie blowing wind. A ray of sunlight passing through this hole falls on the floor across which there is a long marble strip, placed north-south, whose center opens up into a circle. The meridian line is inserted along this marble strip and the signs of the zodiac are cut into it at those positions where the ray of sunlight coming through the hole falls at noon on the day when the sun enters the respective constellations. One of these positions, of course, coincides with the vernal and autumnal equinoxes. By observing with this meridian Gregory XIII is said to have come to the personal realization of the absolute need for the reform of the calendar.[3]

We can infer that the church in 1576 was not antiscience. The recording of time, the position of the sun, and the reform of the calendar we use today (the Gregorian calendar, named in honor of Pope Gregory XIII) were obviously no threats to the authority of the church, to its power, or to the control of the minds of its followers.

The centuries that followed saw a full-fledged observatory at the Vatican in Rome. In 1888 instrumentation at the Vatican Observatory included Merz refractors[4] of aperture 10.2 cm and 10.6 cm and instruments for meteorological observations and for measures of terrestrial magnetism.[5] A rotating dome of 3.5 meters was installed on the Tower of the Winds. This was the first of four domes to be erected in the Vatican. More recently, the Vatican Observatory has spread its wings to Mount Graham in Arizona, with its 1.8-m telescope set in an environment with typically pristine skies.[6] The observatory headquarters remain at Castel Gandolfo near Rome.

Maffeo explains what rivets the eye as one approaches Castel Gandolfo:

> The thousands of visitors, pilgrims and tourists, who come every year to Castel Gandolfo [the Papal summer residence], to take part

---
3. Sabino Maffeo, *The Vatican Observatory: In the Service of Nine Popes*, trans. George V. Coyne, 2nd ed. (Vatican City: Vatican Observatory Publications, 2001), 4.
4. *Refracting* (as opposed to *reflecting*) telescopes use lenses only.
5. Maffeo, *Vatican Observatory*, 34.
6. At the heart of this telescope is a 1.8-m f/1.0 honeycombed-construction (borosilicate) primary mirror.

in the papal audiences in the courtyard of the Apostolic Palace cannot help but notice the domes of the Specola Vaticana [the Vatican Observatory] situated on the top terrace of this ancient building and those half hidden amidst the trees in the Villa Barberini.[7]

The existence of a modern observatory in such a setting causes one to reflect. Many questions are asked of the astronomers of the Specola. Why a papal observatory?

In a section titled "Unmasking the Prejudices," Maffeo hits the nail on the head:

> Above all we must remember that a century ago, at the time when the Specola was re-founded, the dominant culture in Italy and abroad was such that *no occasion was lost to throw weighty accusations at the Church of being obscurantist and closed to scientific progress.* It was precisely to neutralize these accusations that Pope Leo XIII wished to give new life to the Specola, so that "everyone might clearly see that the Church and her pastors are not opposed to true and solid science, whether divine or human, but that they embrace it, encourage it, and promote it with the fullest possible dedication."[8]

The underlying agenda of restoring public trust by erecting an observatory at the Vatican (with fully functional telescopes) is here clearly stated, by Pope Leo XIII himself, in 1891.

Concerning the Vatican Observatory, several commemorative medals have appeared over time. In 1891, for example, a commemorative medal bears this inscription on the front:

> LEO. XIII. PONT. MAX. AN. XIV. ["Leo XIII, Supreme Pontiff, Fourteenth Year"]

On the other side, we read,

> REI. ASTRONOMICAE. HONOR. IN. VATICANO. INSTAVRATVS. ET. AVCTVS. ["Astronomy, established and honoured in the Vatican"][9]

---

7. Maffeo, *Vatican Observatory*, 287.
8. Maffeo, *Vatican Observatory*, 287–88, quoting Pope Leo XIII; italics added.
9. Maffeo, *Vatican Observatory*, 360.

Trust was certainly on the agenda of Pope Leo XIII in 1891. These efforts may have won back trust in the minds of some, but one cannot find any monument or bust of the father of modern science, Galileo Galilei, at the Vatican. Deep in the Vatican archives lie the documents of the Inquisition condemning Galileo, perhaps rather too close to home. His trial and his sentence contain no honor.

One of the cardinals who was "well aware of the enormous damage that had been done in the world of culture by the condemnation of Galileo" was Cardinal Pietro Maffi, who became archbishop of Pisa in 1903. Here was a cardinal who, in 1889, founded a scientific periodical "to spread knowledge of the physical, mathematical and natural science in Italy."[10]

Maffi was a former president of the Vatican Observatory (Specola Vaticana) during the years 1904–1931 and was a full member of the Italian Meteorological Association, of the Italian Society for the Natural Sciences, of the French Astronomical Society, of the Italian Physical Society, and of the Italian Astronomical Society. His efforts to restore Galileo's legacy were quite admirable, but they fell on deaf ears:

> [Maffi] was well aware of the enormous damage that had been done in the world of culture by the condemnation of Galileo and he was ahead of his times in setting out to raise funds to erect at Pisa [Galileo's birthplace], in the Piazza dei Miracoli, a monument to repay the damage done to this great native son [Galileo]. The project drew public interest both in Italy and abroad—it is 1924—and provoked lively discussions about *whether it was fitting to erect the monument* and, if so, where to put it. Representatives of the Civic Administration of Pisa gave indications that they were inclined to accept the Archbishop's proposal, but then after a "reconsideration" they came up with a clear refusal.[11]

The ending of this tragic story is told by Maffeo:

> And so it was that, due to the anticlerical opposition, the funds collected for the monument were used to produce a bronze group

---

10. Maffeo, *Vatican Observatory*, 96–97.
11. Maffeo, *Vatican Observatory*, 106; italics added.

of angels adoring an ostensorium. One can still admire this piece at the principal altar of the Pisa cathedral.[12]

At the time of Galileo's death in 1642, the Grand Duke of Tuscany wished to have him buried in the Basilica of Santa Croce in Florence. Pope Urban VIII objected, on the grounds that the Catholic Church had denounced Galileo as a suspected heretic because of his claims that the earth revolved around the sun. Galileo was buried twice. In a timeline provided by Maurice A. Finocchiaro, we read, "9 January 1642. Galileo is quietly buried at the Church of Santa Croce in Florence, in an unmarked grave located in an out-of-the-way room behind the sacristy and under the bell tower."[13] It took some one hundred years for Galileo's remains to be interred in a marble sarcophagus located directly across from Michelangelo's monument.

The treatment of Galileo by the church of Rome needed careful repair. Again citing Maffeo, "To win trust in this [scientific] field and to undo the condemnation of Galileo concretely and unequivocally it was necessary to involve the Church in scientific research itself."[14]

The Vatican's involvement in astronomical research continues today at a high scientific level. The Vatican Observatory in Arizona with its 1.8-m advanced-technology telescope is a significant research facility. The Vatican Observatory in Rome continues to organize a well-known series of summer schools for undergraduate and graduate students in astronomy at Castel Gandolfo, and it hosts many international research conferences, some of which we have been privileged to attend. One of their most famous conferences was held in Vatican City in May 1957 on the subject of "stellar populations" in galaxies. This concept was pioneered by Walter Baade of the Mount Wilson Observatories and Palomar Observatories in California during World War II, when the dimmed lights of Los Angeles enabled his groundbreaking observations of individual stars in the nearby Andromeda galaxy. The attendees at this conference included most of the leading luminaries in the subject, including Walter Baade himself, Jan Oort

---

12. Maffeo, *Vatican Observatory*, 106.
13. Maurice A. Finocchiaro, ed. and trans., *The Essential Galileo* (Indianapolis: Hackett, 2008), 24.
14. Maffeo, *Vatican Observatory*, 288.

(Leiden), Bertil Lindblad (Stockholm), Allan Sandage (Pasadena), Martin Schwarzschild (Princeton), and Fred Hoyle (Cambridge). Baade's concept of stellar populations changed astronomy forever. Hoyle's profound summary of the conference was a turning point in crystallizing the subject for the future.

Giorgio de Santillana puts it memorably: Galileo is "more than the creator of an era. He has become . . . the *symbol* of a great adventure like Prometheus, or rather like the Ulysses of Dante and Tennyson."[15]

To study the open books of nature and of Scripture is *the* great adventure, for they encompass the nature of truth.

---

15. Giorgio de Santillana, "Galileo in the Present," in *Homage to Galileo*, ed. Morton F. Kaplon (Cambridge: Massachusetts Institute of Technology Press, 1966), 1. See also G. V. Coyne, M. Heller, and J. Życiński, eds., *The Galileo Affair: A Meeting of Faith and Science* (Vatican City: Specola Vaticana, 1985), 172.

Figure 1 Abell 2218 cluster of galaxies. NASA / STScI / Andrew Fruchter and the ERO Team (Sylvia Baggett, STScI; Richard Hook, ST-ECF; Zoltan Levay, STScI). The massive cluster of galaxies known as Abell 2218 lies in the constellation of Draco and is some two billion light-years away from the earth. The cluster is so massive that its enormous gravitational field deflects light rays passing through it, much as an optical lens bends light to form an image. This phenomenon, called gravitational lensing, magnifies, brightens, and distorts images from faraway objects. The cluster's magnifying powers provide a "zoom lens" for viewing distant galaxies that could not normally be observed with the largest telescopes. The deflection of light rays by gravity is predicted by Einstein's theory of general relativity and produces the arc-shaped patterns found throughout this dramatic picture. All "arcs" are the distorted images of very distant galaxies, which lie five to ten times farther than the lensing cluster Abell 2218 itself. Their light was emitted when the universe was a mere quarter of its present age.

Figure 2 The northern lights. Photograph by David L. Block. Used by permission. The aurora borealis, or northern lights, was photographed from Sommarøy, Norway, at a latitude of only 20 degrees from the North Pole.

Figure 3 Windblown trees in Twistleton Scar in the Yorkshire Dales. Photograph by PhilMacDPhotos/Shutterstock.com. Used by permission. Do the trees move the wind, or does the wind move the trees?

Figure 4 Title page to William Tyndale's 1526 translation of the New Testament. Photograph by David L. Block. Used by permission. We behold two books: the book of Scripture and the book of nature.

Figure 5 Professor Jean Mesnard. Photograph at the Musée Rodin (Rodin Museum) in Paris by David L. Block. Used by permission. The late Professor Jean Mesnard was one of the planet's foremost experts on the God of grace in the writings of Blaise Pascal. Professor Mesnard was a Commander of l'Ordre des Palmes académiques, Commander of the Ordre des Arts et Lettres, Chevalier of the Legion d'honneur, and an elected member of the Académie des Sciences Morales et Politiques. He identified a crucial mistake made by Galileo Galilei.

Figure 6 Composite of the planet Saturn with six of its moons. Photo by NASA. This composite was prepared from an assemblage of images taken by the *Voyager 1* spacecraft during its encounter with Saturn. The moon Dione lies in the forefront with Saturn rising behind. The five other moons depicted here are Tethys, Mimas, Enceladus, Rhea, and, in its distant orbit at top right, Titan. The Voyager Project is managed for NASA by the Jet Propulsion Laboratory, California Institute of Technology, Pasadena, California.

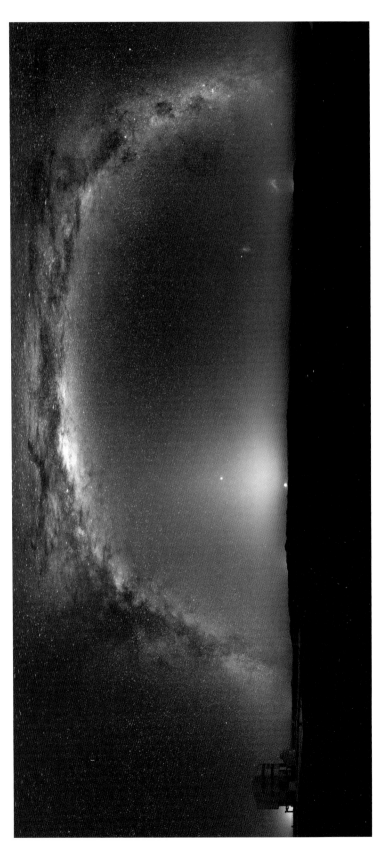

Figure 7 360-degree panorama of the southern sky. Photograph by ESO / H. H. Heyer (http://www.eso.org/public/images/vlt-mw-potw/). Used by permission. The Milky Way arches across this rare 360-degree panorama of the night sky above the Paranal platform, home of the European Southern Observatory's Very Large Telescope (four giant 8.2-meter Unit Telescopes, one of which is shown at left). The photograph combines thirty-seven individual frames, with a total exposure time of about thirty minutes. To the right of the image and below the arc of the Milky Way lie two of our galactic neighbors, the Small and Large Magellanic Clouds.

Figure :: Rosette Nebula. Photograph by David L. Block using the 1-meter Schmidt telescope at the European Southern Observatory at La Silla in Chile. Used by permission. The Rosette Nebula is an example of a stellar maternity ward, where the formation of stars has been triggered in giant clouds of cosmic dust and gas. The Rosette Nebula lies more than five thousand light years away, in the constellation of Monoceros in our Milky Way galaxy.

Figure 9 Spiral galaxy Messier 83. Photograph by ESO (http://www.eso.org/public/images/eso0825a/). Used by permission. This dramatic image of the galaxy Messier 83 was captured by the Wide Field Imager at the European Southern Observatory's La Silla site. Messier 83 lies approximately fifteen million light-years away in the southern constellation of Hydra. In some respects, Messier 83 is rather similar to our own spiral galaxy. Both the Milky Way and Messier 83 contain a "bar," referring to the rather straight or linear conglomeration of stars seen extending on either side of the center.

Figure 10 Jupiter and its four "Medicean" moons. Photograph by NASA/JPL. Galileo Galilei discovered these four moons. Reddish Io is at the upper left, Europa in the center, Ganymede at the bottom, and Callisto at the lower right. The images were secured by the *Voyager 1* spacecraft and assembled into this collage, which is not to scale. The Voyager Project is managed for NASA by the Jet Propulsion Laboratory, California Institute of Technology, Pasadena, California.

Figure 11 The villa Il Gioiello, Galileo's last home. Photograph by David L. Block. Used by permission. Galileo died in this villa under house arrest following his sentence by the Inquisition in Rome.

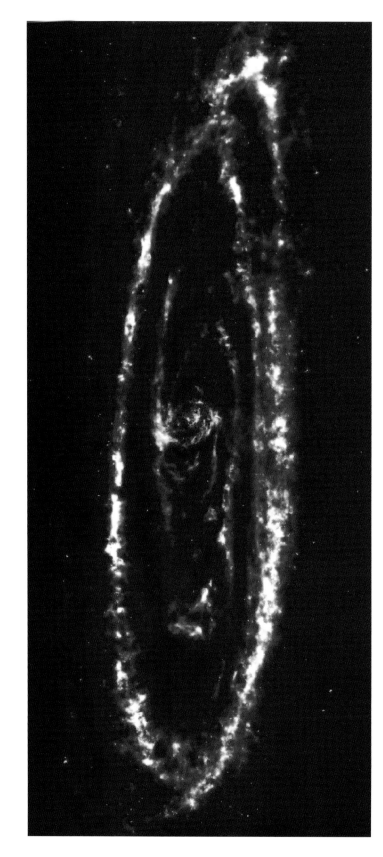

Figure 12 Andromeda spiral galaxy. Photograph by NASA / JPL-CALTECH / David L. Block. Cosmic Dust Laboratory, University of the Witwatersrand, South Africa. Used by permission. The Andromeda spiral galaxy, approximately 2.5 million light years distant, is imaged in the near-infrared region of the spectrum, through the eyes of the Spitzer Space Telescope. Two prominent rings of star formation ("rings of fire") are seen; the diameter of the outer ring spans over sixty thousand light years.

Figure 13 Cosmologist Allan Sandage. Photograph by Robert Groess. Used by permission. Cosmologist Allan Sandage, right, is standing next to David Block, left, at the Athenaeum in Pasadena, California, beneath a painting by S. Seymour Thomas. Displayed in the painting are Nobel laureate Robert A. Millikan (center), former MIT President Arthur A. Noyes (left) and astronomer George Ellery Hale (right), after whom the giant 5-meter Hale telescope at Mount Palomar is named.

Figure 14 Ice glass. Produced by master glassmaker Andrea Zilio and his assistants in Murano, Venice. Photograph by David L. Block. Used by permission. In his *Sidereus Nuncius*, Galileo compares the appearance of the moon through his spyglasses to that of cracked or "ice glass." Seen here is a miniature moon (diameter 18 cm) of cracked glass. Galileo's powerful analogy of the unending chain of peaks, valleys, and craters on the moon with cracked glass on the earth comes to life in this work.

Figure 15 *Portrait of Galileo Galilei*, by Giacomo Ciesa, 1772–1773. Photo courtesy of Museo *La Specola*, INAF–Astronomical Observatory of Padova.

PART 3

PERSONAL EXPERIENCES OF GRACE

12

Grace in the Life of Blaise Pascal

Grace is not religion. Grace is a person: God incarnate. One sees God's grace wonderfully operating in the life of Blaise Pascal (1623–1662).

Any serious student of mathematics, physics, and computer science will have encountered the name of the legendary French scientist Blaise Pascal. W. W. Rouse Ball sums up his most remarkable early achievements as follows:

> At the age of fourteen he was admitted to the weekly meetings of Roberval, Mersenne, Mydorge, and other French geometricians; from which, ultimately, the French Academy sprung. At sixteen Pascal wrote an essay on conic sections; and in 1641, at the age of eighteen, he constructed the first arithmetical machine, an instrument which, eight years later, he further improved. His correspondence with Fermat about this time shews that he was then turning his attention to analytical geometry and physics. He repeated Torricelli's experiments, by which the pressure of the atmosphere could be estimated as a weight, and he confirmed his theory of the cause of barometrical variations by obtaining at the same instant readings at different altitudes on the hill of Puy-de-Dôme.[1]

---

1. W. W. Rouse Ball, "A Short Account of the History of Mathematics," trans. D. R. Wilkins, 4th ed. (London: Macmillan, 1908), 282–88, http://www.pascal-central.com/blaise.html.

Professor H. Wildon Carr penned these words in 1923, the tercentenary of Pascal's birth:

> Descartes was shown the Treatise on Conic Sections which Pascal composed when sixteen, and refused to believe in its originality....
>
> In science and philosophy he (Pascal) showed an intellectual power and incentive which places him on a level with Descartes and Galileo, yet he stands alone, grand but solitary, in the great intellectual movement of humanity.[2]

What a remarkable duo—Galileo and Pascal. Two champions: Galileo, father of modern astronomy, and Pascal, one of the greatest Christian apologists of all time. Pascal understood that, in the nature of truth, scholastic rationalism has its limits—there is the world of the heart. Grace transforms the heart. Professor Jean Mesnard (see fig. 5) elaborates:

> I would say that for Pascal (and this is my personal opinion), when we discover faith, we feel a certain enthusiasm, which etymologically speaks of being possessed by God. Enthusiasm is being possessed by God. When Pascal experienced his conversion, he was already convinced by rational proofs, but he considered that this wasn't enough. There must be a move of the heart, and that is enthusiasm. In the "Memorial" he says, "Joy, joy, joy, tears of joy." Joy is precisely a form of enthusiasm.[3]

Pascal had a cataclysmic encounter with Jesus on November 23, 1654. It is referred to as Pascal's "night of fire." His "Memorial" (alluded to earlier) records his encounter with God on that unforgettable night. The lines written by Pascal are his record of God's revelation for approximately two full hours. Pascal always carried these lines with him: the script was found in the lining of his coat after his death at age thirty-nine. It would be a grave injustice to dissect the "Memorial" and to extract only certain sentences. The complete "Memorial" reads as follows:

---

2. H. Wildon Carr, "The Tercentenary of Blaise Pascal," *Nature* 111, no. 2798 (1923): 816.
3. A conversation between David Block and Jean Mesnard, Paris, March 1, 2014. Translated from French into English by Sarah Frewen-Lord.

This year of Grace 1654,
Monday, November 23rd, day of Saint Clement, pope
and martyr, and others in the martyrology,
Eve of Saint Chrysogonus, martyr, and others;
From about half past ten at night, to
about half after midnight,
Fire.
God of Abraham, God of Isaac, God of Jacob,
Not of the philosophers and the wise.
Security, security. Feeling, joy, peace.
God of Jesus Christ
Deum meum et Deum vestrum.
Thy God shall be my God.
Forgetfulness of the world and of all save God.
He can be found only in the ways taught
in the Gospel.
Greatness of the human soul.
O righteous Father, the world hath not known thee,
but I have known thee.
Joy, joy, joy, tears of joy.
I have separated myself from him.
Dereliquerunt me fontem aquæ vivæ.
My God, why hast thou forsaken me? . . .
That I be not separated from thee eternally.
This is life eternal: That they might know thee
the only true God, and him whom thou hast sent, Jesus Christ,
Jesus Christ,
Jesus Christ.
I have separated myself from him; I have fled, renounced,
    crucified him.
May I never be separated from him.
He maintains himself in me only in the ways taught
in the Gospel.
Renunciation total and sweet.
etc.[4]

---

4. Blaise Pascal, "Pascal's Profession of Faith," in *The Thoughts of Blaise Pascal*, trans. C. Kegan Paul (London: Kegan Paul, Trench, 1885), 2, http://www.gutenberg.org/files/46921/46921-h/46921-h.htm.

Pascal encountered the God of purity—of fire. At once, Pascal recorded that he was speaking of the Jewish God: "God of Abraham, God of Isaac, God of Jacob."

Pascal's experience did not rely on philosophy or on scholarly books: it was not a work of the intellect but a work of grace. Pascal's *spiritual* eyes were opened, as in a flash. He spoke of the *greatness of the human soul*. His soul was ignited. Pascal met Jesus—"God of Jesus Christ"—as if face-to-face. To *know* God is to know him with certainty. Pascal affirmed, "Security, security. Feeling, joy, peace." He *knew* that he *knew* that he *knew*, by an infusion of revelation and grace, that he had met the living God of Abraham, Isaac, and Jacob—the "God of Jesus Christ."

Pascal's "Memorial" reechoed the words of our Lord in John 17:3: "And this is eternal life, that they may know You, the only true God, and Jesus Christ whom You have sent" (NKJV). Pascal's "night of fire" in 1654 blazes with an eternal truth: "O righteous Father, the world has not *known* you, but I have *known* you" (John 17:25 NKJV).

Such was the nature of truth in the life of Blaise Pascal.

13

Grace alongside a Telescope in South Africa

Because salvation is a central theme found in Galileo's *Letter to the Grand Duchess Christina* and is also at the heart of the ministry of Jesus, allow us to expand on what is meant by *sin* and by *salvation*. Sin is the act of defying God's will. Salvation is the saving of the soul from sin and its consequences and the reconciliation of God with his wayward people. To personalize this, we indeed, like sheep, do spiritually go astray, but the eyes of the Shepherd are always on us.

A most insightful discussion about sin and guile and deceit comes from a few lines of a poem written by Galileo, "Against Wearing the Gown":

> Those who wish to know what abstinence is
> should first consider the nature of indulgence at Carnival time,
> and define the difference between them.
> And if they wish to learn about sin,
> should see if a priest will assign them a penance.
> And if you want to acquaint yourself with rogues,
> base men of little distinction,
> just get to know the priests and friars
> who appear all goodness and devotion.
> This is how to get to the bottom
> and unravel our problem.

> It can clearly be seen that the sole cause
> of all guile and deceit
> is to go forth always in fine attire.
> Another and no lesser curse
> derives for us from this evil seedbed
> which brings confusion to the world:
> and that is the privilege and prestige
> the wearing of white, black or brown habits confers
> which creates discrimination among Christians.[1]

Privilege. Prestige. Outward appearance. What about our inward state—the frightful condition of our hearts? A riveting discussion concerning the physical world, as contrasted to the world of the Spirit, is found when Nicodemus, a leader of the Jews, comes to Jesus by night:

> There was a man of the Pharisees named Nicodemus, a ruler of the Jews. This man came to Jesus by night and said to Him, "Rabbi, we know that You are a teacher come from God; for no one can do these signs that You do unless God is with him."
>
> Jesus answered and said to him, "Most assuredly, I say to you, unless one is born again, he cannot see the kingdom of God."
>
> Nicodemus said to Him, "How can a man be born when he is old? Can he enter a second time into his mother's womb and be born?"
>
> Jesus answered, "Most assuredly, I say to you, unless one is born of water and the Spirit, he cannot enter the kingdom of God. That which is born of the flesh is flesh, and that which is born of the Spirit is spirit. Do not marvel that I said to you, 'You must be born again.' The wind blows where it wishes, and you hear the sound of it, but cannot tell where it comes from and where it goes. So is everyone who is born of the Spirit." (John 3:1–8 NKJV)

Walking in his realm is likened to walking in the wind. As we have noted earlier, we cannot see the wind with our physical eyes; we only see the effects of the wind blowing.

---

1. The translation is by Anna Teicher of Cambridge and is used with permission. The original Italian title of the poem is "Capitolo contro il portar la toga" and the poem itself is found in Galileo, *Le Opere di Galileo Galilei*, ed. Antonio Favaro et al., vol. 9, *Scritti Letterari* (Florence: G. Barbèra, 1968), 213–23.

We either see and experience the grace, love, and forgiveness of God, or we are *blind* to it, as emphasized earlier by Scot Bontrager. Only Jesus can make the spiritually sightless to *see* and *know* his presence.

Here is a brief but eloquent description of salvation by A. W. Tozer (1897–1963), an American Christian preacher and author: "Salvation is from our side a choice, from the divine side it is a seizing upon, an apprehending, a conquest by the Most High God."[2]

D. L. Moody (1837–1899), an American evangelist, who founded the school that would become Moody Bible Institute, penned these words:

> The thief had nails through both hands, so that he could not work; and a nail through each foot, so that he could not run errands for the Lord; he could not lift a hand or a foot toward his salvation, and yet Christ offered him the gift of God; and he took it. Christ threw him a passport, and took him into Paradise.[3]

Coming to faith in Jesus is a deeply personal experience. Grace transcends time and circumstance. The Spirit of God moves uniquely upon different individuals, but the end result—salvation—is always the same. As we read in the Gospels, "The wind blows [breathes] where it wishes, and you hear the sound of it, but cannot tell where it comes from and where it goes. So is everyone who is born of the Spirit" (John 3:8 NKJV). We are not dealing with a distant God who sets his universe into motion and who subsequently leaves it alone. God is active in his creation, as is graciously revealed in the Scriptures. Each person has his or her own experience of coming to faith in Jesus. In what follows, we both share some thoughts about our own faith experiences.

David Block's Story

I grew up as a Jewish boy in Krugersdorp, South Africa, and I would be enthralled, sitting in shul (synagogue), as our learned rabbi expounded

---

2. A. W. Tozer, *The A. W. Tozer Bible: King James Version* (Peabody, MA: Hendrickson, 2012), 168.
3. D. L. Moody, *"Let the Wicked Forsake His Way": An Address* (London: Morgan and Scott, 1881), 12.

how God was a *personal* God—God would speak to Moses, to Abraham, to Isaac, to Jacob—and to many others. I pondered much how I fit into all this. By the time I entered university, I became deeply concerned that I had *no assurance* that God was indeed a *personal* God. I did know that he was a *historical* God and that he did deliver our people from the hands of Pharaoh. But he seemed so far removed from me in Krugersdorp. Where was the personality and the vibrancy of a God who could speak *to me*?

While a student at the University of the Witwatersrand, Johannesburg, I began studying for my bachelor of science degree in applied mathematics and in computer science. I became friendly with the late Lewis Hurst, then a professor of genetics and medicine. He had a great interest in astronomy, and for many hours we would discuss the complexities of the cosmos. I would delight in explaining to him fundamentals in astronomy, such as black holes and quasars.

Intellectually, I was satisfied. I became fascinated by the elegance of the mathematical formulation of general relativity, and I submitted my first research paper on that theme to the Royal Astronomical Society in 1973 (at age nineteen); it was published by that society one year later.[4] How humbling it was for me to receive requests from observatories and universities for reprints or printed copies of that research paper; all the requests were invariably addressed to "Professor David Block"—little did anyone know my age and that I was merely a "Mr."!

I well remember attending a meeting of the Royal Astronomical Society in London during that period; Stephen Hawking was in attendance. Intellectually, I was so stimulated, but questions within gripped me. Inwardly, *something*, or Someone, was missing.

Back in South Africa, my friendship with Professor Hurst grew—and I started sharing my thoughts and feelings about the cosmos with him. "The universe is so beautiful," I proclaimed, "both visually and mathematically." Professor Hurst listened intently. That there was a Master Artist (Jewish tradition affirmed that he actually existed) continued to resonate with me, but where was *my God*? Was the incomprehensible yet comprehensible world that I studied indeed one of purpose?

---

4. David Lazar Block, "General Relativity, and Its Applications to Selected Astrophysical and Cosmological Topics," *Quarterly Journal of the Royal Astronomical Society* 15 (1974): 264–91.

To be brutally honest, I did not *know* God.

"What concerns me, deeply so, is that the universe is so large, so immense. Is physical reality the sum total of our existence?"—a question I often reflected on as a young university student in South Africa.

I shared further doubts with Professor Hurst: "Are we, as Shakespeare said, just a fleeting shadow to appear and then disappear? What is our purpose for living? What is the *raison d'etre* for being here? Was it possible to ever have a personal encounter with the Creator of the cosmos?

"There is an answer to all the questions you are asking," Professor Hurst responded. "I am well aware that you come from an Orthodox Jewish family . . . but would you be willing to meet with a dear friend of mine, the Reverend John Spyker?" His voice resonated with tones of gravity.

My Jewish parents had taught me to seek answers where they may be found—and so I consented to meet with this Christian minister. Strangely, in my heart, when I aimed my first telescope at Saturn (see fig. 6)—when I actually viewed Saturn in all its majesty and splendor, with its tilted system of rings—I just *knew* that there was and is a Great Designer. In fact, somehow I knew that there must be a personal God, but I had not yet *experienced* his still small voice of forgiveness and of reassurance in my heart.

"The heart has its reasons, which reason does not know," penned the brilliant mathematician Blaise Pascal.[5] Reflecting on those moments now, I realize that they were moments infused by the grace of God. It was as if Jesus were sitting at my table (in my case, drawing alongside me at my very first telescope), just as he did at Emmaus.

The voice of the Reverend John Spyker always commanded attention, for he spoke with authority. He took his Bible in his hands and turned the pages to the New Testament—in particular, to the apostle Paul's letter to the Romans. In 9:33, Paul affirms that Y'shua (Jesus) is a stumbling stone to my Jewish people but that those who freely choose to believe in Y'shua will never be ashamed. The relevant section in the King James Version reads as follows: "As it is written, Behold, I lay in Sion a stumblingstone and rock of offense: and whosoever

---

5. From Blaise Pascal, *Pensées* (1670), in Blaise Pascal, *Thoughts*, trans. W. F. Trotter, in vol. 48 of *The Harvard Classics*, ed. Charles W. Eliot (New York: Collier, 1910), 22.

believeth on him shall not be ashamed." For those who prefer a more modern rendition of this verse, *The Message* by Eugene Peterson reads, "Careful! I've put a huge stone on the road to Mount Zion, a stone you can't get around. But the stone is me! If you're looking for me, you'll find me on the way, not in the way."

By divine grace, it all *suddenly* (instantaneously) became very clear to me. Y'shua was the stumbling stone. *My* stumbling stone. Jesus had fulfilled all the messianic prophecies in the Hebrew Scriptures (such as where the Jewish Messiah would be born, how he was to die, and much besides). While many of my Jewish people were still awaiting the Messiah, I suddenly knew that I knew that I knew that Jesus is the living Messiah! All I had to do that day was respond to his grace. I immediately asked the Reverend John Spyker to pray for me. I surrendered my heart and my reason to him that day. The Spirit of Jesus infused every cell of my being. That was in October 1976. I was twenty-two.

It might seem strange to some that a young scientist who was an orthodox Jew should come to faith in Jesus. But faith is never a leap in the dark. As we read in Scripture, faith is based on evidence. In my study of astronomy, I was immersed in the *city of God* ever since receiving my first telescope in high school in 1971 at age seventeen. At that time, while I was still a young man, looking at Saturn was a moment of deep revelation to me, as noted earlier. God had ignited my spiritual candle, and the spiritual seeds implanted took a few years to grow to fruition, from 1971 to 1976. In his essay titled *Nature*, Ralph Waldo Emerson penned these words:

> If the stars should appear one night in a thousand years, how would men believe and adore; and preserve for many generations the remembrance of the *city of God* which had been shown! But every night come out these preachers of beauty, and light the universe with their admonishing smile.[6]

Psalm 19 affirms that the heavens declare the glory of God. From the city of God that I had observed through my telescope beginning in 1971, the year 1976 marked a meeting of macrocosm with microcosm: the Creator, by his grace, had revealed himself to me.

---

6. R. W. Emerson, *Nature* (Boston: James Munroe, 1836), 9–10; italics added.

The sense of awe cannot be verbally described. The heavens must be *observed*. When he observed the Milky Way, the great astronomer Edward Emerson Barnard[7] exclaimed,

> Sweep on through glittering star fields and long for endless night! More nebulae, more stars. Here a bright and beautiful star overpowering in its brilliancy, and there close to it a tiny point of light seen with the greatest difficulty, a large star and its companion. How plentiful the stars now appear. Each sweep increases their number. The field is sprinkled with them, and now we suddenly sweep into myriads and swarms of glittering, sparkling points of brilliancy—we have entered the Milky Way. We are in the midst of millions and millions of suns—we are in the jewel house of the Maker, and our soul mounts up, up to that wonderful Creator, and we adore the hand that scattered the jewels of heaven so lavishly in this one vast region. No pen can describe the wonderful scene that the swinging tube reveals as it sweeps among that vast array of suns.[8]

Such is the city of God.

I have always been vocal and honest about my encounter with God, both in my public meetings and on radio and television. This has led to interesting discussions at times. I recall a fascinating visit many years ago, by a senior academic. The academic had gone to considerable lengths to applaud my scientific publications over the years in *Nature*. I felt grateful for my visitor's acknowledgments.

I knew that it was *God* who had been inextricably involved in my research. He had given me the privilege of working, as a team, with some of the most gifted astronomers alive. At the very beginnings of my career, as I studied the shapes (morphologies) of galaxies, with their dark and pervasive clouds of dust, God had highlighted one specific verse to me: "And I will give thee the treasures of darkness, and hidden riches of secret places, that thou mayest know that I, the LORD, which call thee by thy name, am the God of Israel" (Isa. 45:3

---

7. It was Edward Emerson Barnard who added a fifth satellite to the four satellites of Jupiter discovered by Galileo in January 1610. Barnard discovered Amalthea on September 9, 1892, using the 36-inch (91-cm) refractor telescope at the Lick Observatory in California.

8. Quoted in William Sheehan, *The Immortal Fire Within: The Life and Work of Edward Emerson Barnard* (Cambridge: Cambridge University Press, 1995), 50.

KJV). A more recent translation might be helpful: "And I will give you treasures hidden in the darkness, secret riches; and you will know that I am doing this—I, the Lord, the God of Israel, the one who calls you by your name" (TLB).

It was God speaking *to me* through the Scriptures by means of his Holy Spirit.[9] I have publicly shared this verse with audiences on innumerable occasions. God has faithfully given us *treasures of darkness*! Clouds of cosmic dust are indeed *dark* to the naked eye. This is a consequence of such clouds both dimming and reddening the light from more distant, background stars.[10]

Back to my visitor. This scholar and I were about to bid farewell when the discussion suddenly changed course. There was an urgency in voice tone: "We are so proud of your scientific achievements, David, but could you please *shut up* about God." God! My God!

Now one trait of academics is invariably that we are readers, whether of books or research essays in journals or newspaper articles or much more besides. Were there, perhaps, any specific books on my visitor's list of favorites? My visitor informed me that it was Richard Dawkins's *The God Delusion*, which was then rather hot off the press. Herein lies the rub: Why is one academic free to express his personal belief that God is a delusion, while another is being implored to "*shut up* about *God*"? Why should *I* shut up about the nature of truth? As I type, the above verse from Isaiah still resonates so richly within my being. Some forty years following that personal encounter with God in 1976, I remain enthralled by the grace and love of Jesus, the luminous figure of the Nazarene.

As a teenager, standing alongside my telescope in Krugersdorp, I was totally unaware of three prongs of the scientific fork:

1. The agenda of some people who would thrive seeing one give up, rather than look up—in other words, an agenda to weaken resolve of one's will
2. The attempt to torpedo careers

---

9. A *rhema* is a verse (or portion of Scripture) that the Holy Spirit brings to our attention. We may need wisdom for a current situation we may be facing, or we may need direction (in my case, direction into which field of research God wanted me to undertake). The Holy Spirit guides us (James 1:5).
10. The appropriate astrophysical terminology would be *interstellar extinction*.

3. A world of disguise and lack of integrity, wherein some persons wear two faces

And yet, all three of these challenges were experienced by Galileo Galilei himself—and I, too, like multitudes of others, could learn from him.[11]

The first item: Multitudes have been riveted by the story of Galileo and attracted to it like a magnet. To me, the Galileo story was an attempt by one of the most powerful institutions in the world to try and break the will of Galileo. George Coyne, citing Richard Blackwell,[12] hits the nail on the head: "Thus, as Blackwell so clearly puts it, the abjuration forced on Galileo in 1633 'was intended to bend—or break—his will rather than his reason.'"[13]

In his foreword to Galileo's *Dialogue concerning Two Chief World Systems*, Albert Einstein also emphasizes this point, most eloquently so. Speaking of Galileo, he says,

> A man is here revealed who possesses *the passionate will*, the intelligence and the courage to stand up as a representative of rational thinking against the host of those who, relying on the ignorance of the people and the indolence of teachers in priest's and scholar's garb, maintain and defend their positions of authority.[14]

The second item: Since I have been immersed in the life of Galileo for several years, the key word that always comes to my mind with this affair is *disguise*. The Aristotelian academics of his day proudly donned their togas or scholarly gowns, but behind many of those gowns lay pride and prejudice. Academics of our day may, too, proudly don their gowns—especially at graduation ceremonies—but beware those who try to torpedo a career.

---

11. Allow me to emphasize that these are my personal observations as a Jewish Christian and as a professional astronomer. I take great pains to emphasize that I am neither a trained historian of science nor a philosopher of science.

12. Richard J. Blackwell, "Could There Be Another Galileo Case?," in *The Cambridge Companion to Galileo*," ed. Peter Machamer (Cambridge: Cambridge University Press, 1998), 348–66.

13. George V. Coyne, "The Church's Most Recent Attempt to Dispel the Galileo Myth," in *The Church and Galileo*, ed. Ernan McMullin (Notre Dame, IN: University of Notre Dame Press, 2005), 340–59. Also see Blackwell, "Another Galileo Case?," 348–66.

14. Galileo Galilei, *Dialogue concerning Two Chief World Systems, Ptolemaic and Copernican*, trans. Stillman Drake, 2nd ed. (Berkeley: University of California Press, 1967), vii, italics added.

Einstein speaks of garb or dress, and so did Galileo (see fig. 15). In his poem cited above, Galileo alludes to the difference between appearance and reality, of wearing the robes of a scientist but not really being one at all. He speaks, tongue in cheek, about those parading as priests but not really being priests at all. The story of Galileo is a crisis of power but not only that. On a personal level, it is the robbing of glory. In 1642, the father of modern science was buried in an unmarked grave. In the mind of Robert Frost, the robbing of glory is the worst of all crimes, worse than the robbing of a grave.

Imagine the joy in Galileo's day had he thrown away his passion for scientific truth to follow the herd. How courageous of Galileo never to embrace "the indolence of teachers in priest's and scholar's garb, [to] maintain and defend their positions of authority," to quote Einstein again.

Secular scientists expressing a distaste for those who profess a personal relationship with God are not rare, and some are very influential in the career paths of others. All is not always transparent in the agendas of certain scientists, and all is not always transparent in the agendas of certain theologians either.

Galileo's career suffered greatly at the hands of his opponents. This was recognized by Pope John Paul II, who said, "Galileo had much to suffer . . . from the men and agencies of the Church."[15] As time progressed, however, John Paul II's Galileo Commission concluded that "in the end it appears that no one was responsible for Galileo's sufferings."[16] Father George Coyne, a former director of the Vatican Observatory, was the senior astronomer of the Galileo Commission, and he bravely wrote a disturbing evaluation—one of gravity and concern—regarding the retreat from what Pope John Paul II had earlier said.[17]

A third item: There are many people with two faces in this world. As I type this paragraph, my reading of Scripture this week focused on the book of Esther and the Jewish festival of Purim. As a young Jewish boy, I vividly recall that joyous festival, with its gragers (handheld

---

15. Pope John Paul II, speaking on November 10, 1979, quoted in Coyne, "Church's Most Recent Attempt," 352.
16. Coyne, "Church's Most Recent Attempt," 353.
17. Coyne, "Church's Most Recent Attempt," 353.

devices that make considerable noise) and the eating of pastries known as hamantaschen, characterized by a triangular shape: Haman's ears, we were told. The story of Purim revolves around a man, Haman, who had two faces. To his beloved King Ahasuerus, he presented one face. To the Jew, face number two. Haman had been advanced (promoted) above all the princes of the kingdom (Est. 3:1). But Haman was determined to exterminate all Jews living in the kingdom of Ahasuerus (3:8–9).

Enter Esther (or Hadassah), a beautiful Jewish woman whom King Ahasuerus chose to be his queen. Alerted to the danger facing her people, she would be God's hand in turning the tide. Queen Esther held two banquets for King Ahasuerus and Haman; Haman was honored and delighted (face number one) to be invited to both. But Esther informed the king at the second banquet of her true identity—she was a Jew (7:4). But there is more. She also revealed the second face of Haman—Haman's plot against our Jewish people to exterminate them. The secret was out. The outcome of the story was that the king ordered Haman to be hung on the very gallows that Haman had constructed for the execution of a Jew—Esther's uncle Mordechai (7:10).

The moral of the story in my own testimony is that I have faced influential Hamans in my life, but my personal experience has been that, by God's grace, he has always set a table before me in the presence of my enemies (Ps. 23:5). My motto is simple: never give up, look up, and try again. Others can try to stop up our wells of progress, just as in days of old: "Now the Philistines had stopped up all the wells which his father's servants had dug in the days of Abraham his father, and they had filled them with earth" (Gen. 26:15 NKJV). But God is a master of unstopping wells (John 4:14).

Many years have transpired since I first stood alongside my first telescope in Krugersdorp. Those were the days of nondigital photography. Now all is digital. However, every time I look at images of the wonder of God's creation—especially nowadays from telescopes in space transmitting images back to earth on computer screens—he quietly stands next to me. His grace is sufficient.

I conclude my story where I began—with Galileo's poem.

How they make me suffer,
those who go in search of the highest good,
but have so far failed to find it,
because, my brain tells me,
it is not in the place where they are seeking it.[18]

Others have vastly different stories to tell in their testimonies, as did Pascal in his testimony hundreds of years ago, but the incarnation of Jesus of Nazareth remains a central thread.

Perspective from Ken Freeman

My lack of doubt about God's existence is at a personal level, in that I am conscious of his presence every moment of the day. I could no more doubt his existence than doubt my own.

Coda

We see before us two books, as Galileo did: the book of nature (a book of process, unfolding the mysteries of our universe step-by-step according to scientific methodologies) and the Bible, the book of Scripture. The book of Scripture is a book of purpose. The blindness of Galileo's opponents was to force the book of Scripture to say what it does not say.

The nature of truth spans a vast horizon, not being restricted to science or to the book of nature alone.

---

18. The translation from Italian into English is by Anna Teicher of Cambridge and is used with permission.

Appendix

Galileo's *Letter to the Grand Duchess Christina of Tuscany*

Galileo's lengthy letter to the Grand Duchess Christina of Tuscany, dated 1615, now follows. We use an excellent modern translation of the letter, by Mark Davie.[1]

Letter to Her Serene Highness, Madame the Dowager Grand Duchess

A few years ago, as your Highness well knows, I discovered many things in the heavens which had been invisible until this present age. Because of their novelty and because some consequences which follow from them contradict commonly held scientific views, these have provoked not a few professors in the schools against me, as if I had deliberately placed these objects in the sky to cause confusion in the natural sciences. They seem to forget that the increase of known truths, far from diminishing or undermining the sciences, works to stimulate the investigation, development, and strengthening of their various fields. Showing a greater fondness for their own opinions than for the truth, they have sought to deny and disprove these new facts which, if they had considered them carefully, would have been confirmed by the very evidence of their senses. To this end they have put forward various

---

1. Galileo, *Letter to the Grand Duchess Christina*, trans. Mark Davie, in *Galileo: Selected Writings*, trans. William R. Shea and Mark Davie, Oxford World's Classics (New York: Oxford University Press, 2012), 61–94. Used by permission of Oxford University Press.

objections and published writings full of vain arguments and, more seriously, scattered with references to Holy Scripture, taken from passages they have not properly understood and which have no bearing on their argument. They might have avoided this error if they had paid attention to a salutary warning by St Augustine, on the need for caution in coming to firm conclusions about obscure matters which cannot be readily understood by the use of reason alone. Speaking about a certain scientific conclusion concerning the celestial bodies, Augustine writes: "Aware of the restraint that is proper to a devout and serious person, one should not entertain any rash belief about an obscure question. Otherwise, when the truth is known, we might despise it because of our attachment to our error, even though the truth may not be in any way opposed to the sacred writings of the Old or New Testament."

With the passing of time the truths which I first pointed out have become apparent to all, and the truth has exposed the difference in attitude between those who simply and dispassionately were unconvinced of the reality of my discoveries, and those whose incredulity was mixed with some emotional reaction. Those who were expert in astronomy and the natural sciences were convinced by my first announcement, and the doubts of others were gradually allayed unless their scepticism was fed by something other than the unexpected novelty of my discoveries or the fact that they had not had an opportunity to confirm them by their own observations. But there are those whose attachment to their earlier error is compounded by some other imaginary self-interest which makes them hostile, not so much to the discoveries themselves as to their author. Once they can no longer deny the facts they pass over them in silence, and, distracting themselves with fantasies and embittered even more by what has pacified and won over others, they try to condemn me by other means. And indeed, I would not pay any more attention to these than to the other criticisms made against me, which I have never taken seriously as I have always been confident of prevailing in the end, were it not for the fact that these new attacks and calumnies do not stop at questioning the extent of my learning—for which I make no great claims—but try to smear me with accusations of faults which are more abhorrent to me than death itself. Even if those who know me and my accusers know that their accusations are false, I have to defend my reputation in the eyes of everyone else.

These people, persisting in their determination to use all imaginable means to destroy me and my works, know that in my astronomical and philosophical studies I maintain that the Sun remains motionless at the centre of the revolutions of the celestial globes, and that the Earth both turns on its own axis and revolves around the Sun. They know, moreover, that I uphold this position not just by refuting the arguments of Ptolemy and Aristotle, but also by putting forward many arguments to the contrary, in particular, some related to physical effects that are hard to explain in any other way. There are also astronomical arguments depending on many things in my new discoveries in the heavens, which clearly disprove the Ptolemaic system, and perfectly agree with and confirm this alternative position. Perhaps they are dismayed by the fact that other propositions which I have put forward, which differ from those commonly held, have been shown to be true, and so have given up hope of defending themselves by strictly philosophical means; and so they have tried to hide the fallacies in their arguments under the mantle of false religion and by invoking the authority of Holy Scripture, which they have applied with little understanding to refute arguments which they have neither heard nor understood.

They began by doing their best to spread abroad the idea that these propositions are contrary to Holy Scripture, and therefore to be condemned as heretical. Then, realizing how much more readily human nature will embrace a cause which harms their neighbour, however unjustly, than one which justly encourages them, they had no difficulty in finding others who were prepared to declare from the pulpit, with uncharacteristic confidence, that they were indeed to be condemned as heretical. They showed little pity and less consideration for the injury they were doing not just to this teaching and those who follow it, but to mathematics and mathematicians everywhere. Now, as their confidence has grown and they vainly hope that this seed, which first took root in their own insincere minds, will grow and spread its branches up to heaven, they spread the rumour that it is shortly to be condemned as heretical by the supreme authority of the Church. And since they know that such a condemnation would not just undermine these two propositions, but would extend to all the other astronomical and scientific observations and conclusions which are logically linked to them, they try to make their task easier by giving the impression, as far as they can at least among the general public, that this view is new and mine alone. They pretend not to know that its author—or rather the one who revived and confirmed it—was Nicolaus Copernicus, a man who was not just a Catholic but a priest and a canon, and so highly esteemed that he was called to Rome from the furthest reaches of Germany to advise the Lateran Council under Pope Leo X on the revision of the ecclesiastical calendar. At that time the calendar was incorrect simply because they did not know the exact length of the year and the lunar month; so the Bishop of Fossombrone, who was in charge of the revision, commissioned Copernicus to undertake the prolonged study necessary to establish these celestial movements with greater clarity and certainty. Copernicus set about this task and, thanks to a truly Herculean series of labours combined with his great intellect, made such great progress in this science that he was able to establish the period of the celestial movements with such a high degree of precision, that he came to be recognized as the supreme astronomer, and his findings became the basis not just for the regulation of the calendar but for tables showing the movements of all the planets. He set his findings down in six books which he published at the request of the Cardinal of Capua and the Bishop of Kulm; and since he had undertaken the work at the request of the Supreme Pontiff, he dedicated his book, *On the Revolutions of the Heavenly Spheres*, to Pope Leo's successor, Paul III. As soon as the book was printed it was received by Holy Church and was read and studied throughout the world, without anyone expressing the slightest scruples about its content. But now that the soundness of its conclusions is being confirmed by manifest experiments and necessary demonstrations, there are those who, without even having seen the book, want to reward its author for all his labours by having him declared a heretic—and this solely to satisfy the personal grudge they have conceived for no reason against someone whose only connection with Copernicus is to have endorsed his teachings.

So I have concluded that the false accusations which these people so unjustly try to make against me leave me no choice but to justify myself in the eyes of the general public, whose judgement in matters of religion and reputation I have to take very seriously. I shall respond to the arguments which they produce for condemning and banning Copernicus' opinion, and for having it declared not just false but heretical. Under the cloak of pretended religious zeal, they cite the Holy Scriptures to make them serve their hypocritical purposes, claiming to extend, not to say abuse, the authority of the Scriptures in a way which, if I am not mistaken, is contrary to the intention of the biblical writers and of the Fathers of the Church. They would have us, even in purely scientific questions which are not articles of faith, completely abandon the evidence of our senses and of demonstrative arguments because of a verse of Scripture whose real purpose may well be different from the apparent meaning of the words.

I hope that I can demonstrate that I am acting with greater piety and religious zeal than my opponents when I argue, not that Copernicus' book should not be condemned, but that it should not be condemned in the way they have done, without having understood it, listened to its arguments, or even seen it. For he was an author who never wrote about matters of religion or faith, or cited arguments based in any way on the authority of Scripture, which he might have misunderstood; but he always confined himself to scientific conclusions regarding the movements of the heavens, using astronomical and geometrical proofs resting first of all on the experience of the senses and detailed observations. This is not to say that he paid no attention to what Scripture says, but he was quite clear in his mind that if his conclusions were scientifically proven, they could not contradict Scripture if it was properly understood. Hence he wrote, at the end of his dedication of the book to the Supreme Pontiff:

> If there should happen to be any idle prattlers who, even though they are entirely ignorant of mathematics, nonetheless take it on themselves to pass judgement in these matters, and dare to criticize and attack this theory of mine because of some passage of Scripture that they have wrongly twisted to their purpose, it is of no consequence to me and indeed I will condemn their judgement

for its rashness. It is well known that Lactantius, in other respects a famous writer, was a poor mathematician, and shows his childish understanding of the shape of the Earth when he mocks those who said that the Earth has the form of a sphere. So we scholars should not be surprised if we too are sometimes made fun of by such people. Mathematics is written for mathematicians, and if I am not deceived, they will recognize that these labours of mine make a useful contribution to the ecclesiastical state of which Your Holiness now holds the highest office.

Such are the people who are trying to persuade us that an author like Copernicus should be condemned without even being read. To suggest that this would be not just legitimate but laudable, they cite various texts from Scripture, from theologians, and from the Councils of the Church. I revere these and hold them to be of the highest authority, and I would regard it as the height of temerity to contradict them, as long as they are used in conformity with the practice of Holy Church. Equally, I do not believe it is wrong to speak out if there is reason to suspect that someone is citing and using such texts for their own ends in a way which is at odds with the holy will of the Church. So I declare (and I believe that my sincerity will speak for itself) my willingness to submit to removing any errors which, through my ignorance in matters of religion, may be found in this letter. I declare, further, that I have no wish to enter into quarrels with anyone on such matters, even on points which may be disputable. My purpose is only that, if in these reflections which are outside my professional competence there is anything, among whatever errors they may contain, which might prompt others to find something useful to Holy Church in reaching a conclusion on the Copernican system, it may be taken and used in whatever way my superiors may decide. Otherwise let this letter be torn up and burnt, for I have no desire for any gain from it which is not in keeping with Catholic piety. Moreover, although many of the points I shall discuss are things which I have heard with my own ears, I freely grant to whoever said them that they did not say them, if that is what they wish, admitting that I could well have misunderstood them. Hence, let my reply be addressed not to them but to those who do hold the opinion in question.

The reason, then, which they give for condemning the view that the Earth moves and the Sun is stationary, is that there are many places in Holy Writ where we read that the Sun moves and the Earth does not; and since Scripture can never lie or be in error, it necessarily follows that anyone who asserts that the Sun is motionless and the Earth moves must be in error, and such a view must be condemned.

The first thing to be said on this point is that it is entirely pious to state, and prudent to affirm, that Holy Scripture can never lie, provided its true meaning has been grasped. But I do not think it can be denied that the true meaning of Scripture is often hidden and very different from the literal meaning of the words. It follows that when an expositor always insists on the bare literal sense, this error can make Scripture appear to contain not only contradictions and statements which are far removed from the truth, but even grave heresies and blasphemies; for it would mean attributing to God feet and hands and eyes, not to mention corporeal and human affections such as anger, repentance, hatred, and sometimes even forgetfulness of past events and ignorance of the future. So since the biblical writers, inspired by the Holy Spirit, stated these things in this way so as to be comprehensible to the untrained and ignorant, it is necessary for wise expositors to explain their true meaning to those few who deserve to be set apart from the common herd, and to point out the particular reasons why they have been expressed in the terms that they have. This principle is so commonplace among theologians that it would be superfluous to cite any authorities to justify it.

From this it seems reasonable to deduce that whenever Scripture has had occasion to speak about matters of natural science, especially those which are obscure and difficult to understand, it has followed this rule, so as not to cause confusion among the common people and make them more sceptical of its teaching about higher mysteries. As I have said and as is clear to see, Scripture has not hesitated to veil some of its most important statements, attributing to God himself qualities contrary to his very essence, solely in order to be accessible to popular understanding. Who then would be so bold as to insist that it had set this aside and confined itself rigorously to the narrow literal meaning of the words when speaking in passing about the Earth, water, or the Sun or some other part of creation? This is all the more unlikely since what it says about these things has nothing to do with the primary intention of Holy Writ, namely divine worship and the salvation of souls, and matters far removed from the understanding of the masses.

This being the case, it seems to me that the starting-point in disputes concerning problems in natural science should not be the authority of scriptural texts, but the experience of the senses and necessary demonstrations. For while Holy Scripture and nature proceed alike from the divine Word—Scripture as dictated by the Holy Spirit, and nature as the faithful executor of God's commands—it is agreed that Scripture, in order to be understood by the multitude, says many things which are apparently and in the literal sense of the words at variance with absolute truth. Nature, on the other hand, never trangresses the laws to which it is subject, but is inexorable and unchanging, quite indifferent to whether its hidden reasons and ways of working are accessible to human understanding or not. Hence, any effect in nature which the experience of our senses places before our eyes, or to which we are led by necessary demonstrations, should on no account be called into question, much less condemned, because of a passage of Scripture whose words appear to suggest something different. For not every statement of Scripture is bound by such strict rules as every effect of nature, and God is revealed just as excellently in the effects of nature as in the sacred sayings of Scripture. This may be what Tertullian meant when he wrote: "I conclude that knowledge of God is first to be found in nature, and then confirmed in doctrine; in nature through his works, and in doctrine through preaching."

This is not to imply that we should not have the highest regard for the text of Scripture. On the contrary, once we have reached definite conclusions in science we should make use of them as the best means of gaining a true understanding of Scripture, and of searching out the meanings which Scripture necessarily contains, since it is absolutely true and in harmony with demonstrated truth. I believe therefore that the purpose of the authority of Holy Scripture is chiefly to persuade men of those articles and propositions which, being beyond the scope of human reasoning, could not be made credible to us by science or by any other means, but only through the mouth of the Holy Spirit.

What is more, even in matters which are not articles of faith the authority of Scripture should prevail over that of any human writings which are not set out in a demonstrative way, but are simply stated or put forward as probabilities. This should be regarded as right and necessary, to the same extent that divine wisdom surpasses human understanding or conjecture. But I do not consider it necessary to believe that the same God who has endowed us with senses, and with the power of reasoning and intellect, should have chosen to set these aside and to convey to us by some other means those facts which we are capable of finding out by exercising these faculties, so that even in scientific conclusions which the evidence of our senses and necessary demonstrations set before our eyes and minds, we should deny what our senses and reason tell us. Least of all do I think this applies in those sciences of which only a tiny part is to be found in scattered references in Scripture, such as astronomy, of which Scripture contains so little that it does not even mention the planets, apart from the Sun and Moon, and once or twice Venus, under the name of Lucifer. For if the sacred writers had intended to teach the people the order and movements of the heavenly bodies, and that therefore we should learn these things from Scripture, I do not believe that they would have said so little about them—almost nothing compared to the infinite, profound, and wonderful truths which are demonstrated in this science.

Indeed, it is the opinion of the holy and learned Fathers of the Church that the biblical authors not only made no claim to teach us about the structure and movements of the heavens and stars, and their appearance, size and distance, but that they deliberately refrained from doing so, even though these things were perfectly well known to them. In the words of St Augustine:

> It is commonly asked what we have to believe about the form and shape of heaven according to Sacred Scripture. Many engage in lengthy discussions on these matters, but our writers, with their greater prudence, have omitted them. Such subjects are of no profit for those who seek a blessed life, and, what is worse, they take up very precious time that ought to be given to what is spiritually beneficial. What concern is it of mine whether heaven is like a sphere and the Earth is enclosed by it and suspended in the middle

of the universe, or whether heaven is like a disc that covers the Earth on one side? But the credibility of Scripture is at stake, and as I have indicated more than once, there is some danger for a man uninstructed in divine revelation. Discovering something in Scripture or hearing something cited from it that seems to be at variance with the knowledge he has acquired, he may doubt its truth when it offers useful admonitions, narratives, or declarations. Hence, let it be said briefly that concerning the shape of heaven the sacred writers knew the truth, but that the Spirit of God, who spoke through them, did not wish to teach men these facts that would be of no avail for their salvation.

The same lack of interest on the part of the sacred writers in laying down what should be believed about these properties of the celestial bodies is shown again by St Augustine in the next chapter, chapter 10, where on the question of whether the heavens should be deemed to be motionless or in motion, he writes:

> Concerning the heaven, some of the brethren have enquired whether it is stationary or moving. If it is moving, they say, how is it a firmament? And if it is stationary, how do the heavenly bodies that are thought to be fixed in it travel from east to west, the more northerly performing smaller circles near the pole? So heaven is like a sphere, if there is another pole invisible to us, or like a disc, if there is no other axis. My reply is that a great deal of subtle and learned enquiry into these questions would be required to know which of these views is correct, but I have no time to go into these questions and discuss them. Neither have they time, those whom I wish to instruct for their own salvation and for the benefit of the Holy Church.

Coming to the particular point with which we are concerned, if the Holy Spirit has chosen not to teach us whether the heavens move or stand still, or whether they have the shape of a sphere, a disc, or a flat surface, or whether the Earth is in the middle of the heavens or to one side, it necessarily follows that He had no intention of giving us a definite answer to other questions of the same kind. The question of the motion or rest of the Earth and the Sun is so linked to those mentioned above that it cannot be determined one way or the other without first answering these. If the Holy Spirit has deliberately refrained from teaching us such things as not being relevant to his intention—that is, to our salvation— how can it be claimed that taking one or other view on this question is obligatory, and that one view is an article of faith and the other is an error? Is it then possible for an opinion to be heretical, and yet have no relevance to the salvation of souls? Or can it be claimed that the Holy Spirit has chosen not to teach us something which concerns our salvation? I cannot do better here than quote what I have heard said by a very eminent churchman, that the intention of the Holy Spirit is to teach us how one goes to heaven, not how the heaven goes.

Returning to the question of how much weight should be given in questions of natural science to necessary demonstrations and the experience of the senses, and how much these have been regarded as authoritative by learned and holy theologians, the following are two statements among many: "In discussing the teaching of Moses, we should take care to avoid at all costs saying or declaring categorically ourselves anything that goes against what is clear from manifest experience and the reasoning of philosophy or other disciplines. Since any truth always agrees with every other truth, the truth of Holy Scripture cannot contradict the truth of human sciences established through experience and reason." And in Saint Augustine we read: "Anyone who invokes the authority of Scripture in opposition to what is clearly and conclusively established by reason, does not understand what they are doing. What they are opposing to the truth is not the meaning of Scripture, which they have failed to grasp, but their own view, which they have found not in Scripture but in themselves."

This granted, and since, as has been said, two truths can never contradict each other, it is the duty of the wise expositor to seek out the true meanings of Scripture, which undoubtedly will agree with those scientific conclusions which observation and necessary demonstrations have already established as certain. Now the Scriptures, as we have seen, often allow interpretations which differ from the meaning of the words, for the reasons given above. Moreover, there is no guarantee that all interpreters are divinely inspired, for if they were, they would never disagree over the meaning of a given passage. So it would be highly prudent not to allow anyone to use Scripture to uphold as true any scientific conclusion which observation and demonstrative and necessary reasons might at some time show to be false. For who can place limits on the human mind, or claim that we already know all that there is to be known? Will it be those who on other occasions admit, quite rightly, that "What we know is only a tiny part of what we do not know"?

Indeed, since we have it from the mouth of the Holy Spirit that "He has given up the world to disputations, so that no man may find out what God made from the beginning to the end," I do not think we should contradict this by closing the path to free speculation concerning the natural world, as if everything had already been discovered and revealed with absolute certainty. Nor do I think it should be considered presumptuous to challenge opinions which were formerly commonplace, or that anyone should be indignant if someone does not share their opinion on a matter of scientific dispute—least of all in the case of problems which the greatest philosophers have debated for thousands of years, such as the view that the Sun is fixed and the Earth moves. This was the view held by Pythagoras and all his followers, Heraclides of Pontus, and also Plato's teacher Philolaus, and by Plato himself, as Aristotle tells us and as Plutarch confirms in his life of Numa, where he writes that Plato in his old age used to say that it was absurd to believe otherwise. The same view was held by Aristarchus of Samos, as we learn from Archimedes; by Seleucus the mathematician; by the philosopher Nicetas, according to Cicero, and by many others; and it has finally been developed and confirmed with numerous observations and proofs by Nicolaus Copernicus. Seneca too, that most eminent philosopher, exhorts us in his book *On Comets* to make every effort to establish with certainty whether it is in the sky or on Earth that the daily rotation is located.

So perhaps it would be only wise and prudent not to add unnecessarily to the articles concerning salvation and the foundations of the faith whose certainty is immune to valid arguments ever being raised against it. It would be doubly unwise to add to them at the request of those who may or may not be inspired from above, but who clearly lack the intelligence first to understand, and then to discuss, the demonstrations which the exact sciences use to validate their conclusions. If I were allowed to give my opinion, I would go further and say that it might be more appropriate, and more befitting the dignity of Holy Writ, to stop every lightweight popular writer from trying to lend authority to their writings, often based on empty fancies, by quoting verses from Scripture, which they interpret or rather force into saying things which are as far from the true meaning of Scripture as they are near to making complete fools of themselves when they parade their biblical knowledge in this way. I could give many examples of such abuses of Scripture; let two suffice, both relevant to these astronomical questions. The first are the writings which were published attacking the Medicean planets, which I recently discovered, which cited many verses of Scripture to prove that they could not exist. Now that these planets are plain for everyone to see, I would like to know what new interpretations of Scripture those who opposed me can give to justify their foolishness. The other example I would give is the writer who has recently published a book arguing, against astronomers and philosophers, that the Moon shines with its own brightness and does not receive its light from the Sun. He confirms—or rather, he persuades himself that he can confirm—this fanciful idea with various passages of Scripture which he thinks make sense only if his opinion is necessary and true. Yet the natural darkness of the Moon is as plain to see as the brilliance of the Sun.

It is clear then that if the authority of these writers had counted for anything, they would have imposed their faulty understanding of Scripture on others and would have made it obligatory to believe as true propositions which run counter to manifest proof and the evidence of the senses. God forbid that such an abuse should ever gain a foothold, for if it did then all the investigative sciences would very soon have to be forbidden; for since there are always far more men who are incapable of properly understanding Scripture and the other sciences than there are men of understanding, they would indulge in their superficial reading of Scripture and claim the authority to pronounce on every question of natural science, on the basis of some verse which they have misunderstood and taken out of the context in which the sacred writers intended it. And they would overwhelm the small number of those who understand such matters, as they would always have more followers, for people will always prefer to gain a reputation for wisdom without the effort of studying than to wear themselves out labouring tirelessly at rigorous scientific disciplines. So we should give thanks to God that in his kindness he has spared us this fear, by denying such people any authority and ensuring that no weight is given to their shallow writings. Rather, the task of consulting, deciding and legislating on matters of such importance has been entrusted to the wisdom and goodness of prudent Fathers and to the supreme authority of those who, guided by the Holy Spirit, cannot fail to decree wisely. I think that it was against such men that the Church Fathers wrote with well-justified indignation, in particular St Jerome, who says:

> The garrulous old woman, the senile old man, and the long-winded sophist all presume to have their say about Scripture, mangling it and teaching before they have learnt. Some, prompted by pride, bandy fine-sounding words as they hold forth about Holy Writ among ignorant women; others, I am ashamed to say, learn from women what they teach to men, and as if that were not enough, glibly expound to others things that they do not understand themselves. I will not even speak of those of my colleagues who, perhaps having come to the Holy Scriptures after a career in secular letters, gratify people's ears with carefully constructed phrases, and think that whatever they say is the word of God, without bothering to find out what the prophets and apostles taught. They adapt incongruous testimonies to their own purposes, as if it was an admirable rather than a deplorable way of teaching to distort the meaning of Scripture and twist it to their own contradictory ideas.

There are some theologians whom I hold to be men of great learning and sanctity of life, and for whom I therefore have the highest esteem, whom I would not wish to count among such profane writers. But I must confess that I do have some doubts which I would gladly have resolved when I hear that they claim, on the basis of the authority of Scripture, to require others to accept in scientific debates the view which they consider best harmonizes with Scriptural texts, while at the same time not accepting any obligation on their part to answer the reasons or evidence given to the contrary. They explain and justify this position by saying that theology is the queen of sciences, and therefore should on no account stoop to adapt to the teachings of other less exalted sciences which are subordinate to her, but rather that they should defer to her as the supreme ruler, and change their conclusions to conform to the statutes and decrees of theology. They go further, and say that if those who profess a subordinate science reach a conclusion which they consider to be certain, because it can be demonstrated or proved experimentally, but which is contradicted by a conclusion stated in Scripture, then it is up to them to disprove their own demonstrations and expose the fallacies in their own experiments, without bothering the theologians and biblical scholars. For, they say, it is not befitting the dignity of theology to stoop to investigating the weaknesses of its subject sciences; its role is solely to determine the truth of the conclusion, with absolute authority and in the certainty that it cannot err. The scientific conclusions about which they say we should defer to Scripture, without trying to gloss or interpret it in any way other than the literal meaning of the words, are those where Scripture consistently says the same thing and which the Church Fathers all receive and expound in the same way. There are several points about this ruling which I will raise so as to be advised by those who understand these matters better than me, and to whose judgement I submit at all times.

First, I fear there may be some cause for confusion if the pre-eminence which entitles theology to be called the queen of sciences is not clearly defined. It could be because the material taught by all the other sciences is encompassed and demonstrated in theology, but by more comprehensive methods and with more profound learning—in the same way as, for instance, the rules for measuring fields or keeping accounts are contained pre-eminently in arithmetic and Euclid's geometry rather than in the practical methods of surveyors or accountants. Or it could be because the subject matter of theology surpasses in dignity the subject matter of the other sciences, and because it proceeds by more sublime methods. I do not think that theologians who are conversant with the other sciences would claim that theology deserves to be called queen for the first of these reasons, for it is hard to believe that any of them would say that geometry, astronomy, music and medicine are more comprehensively and precisely expounded in Scripture than in the works of Archimedes, Ptolemy, Boethius and Galen. It follows that the regal pre-eminence of theology must be of the second kind, namely on account of its elevated subject matter, its marvellous teaching of divine revelation, which human comprehension could not absorb in any other way, and its supreme concern with how we gain eternal beatitude. And if theology is concerned with the most elevated contemplation of the divine, occupying its regal throne because of its supreme authority, and does not stoop to the baser and more humble concerns of the subordinate sciences but rather, as has been said above, has no interest in them because they do not concern our beatitude, then those who practise and profess it should not claim the authority to lay down the law in fields where they have neither practised nor studied. If they did, they would be like an absolute prince who, knowing he was free to command obedience as he wished, insisted that medical treatment be carried out and buildings be constructed as he dictated even though he was himself neither a doctor nor an architect, thereby causing grave danger to the lives of his unfortunate patients and the evident ruin of his buildings.

Then, to command that professors of astronomy should be responsible for undermining their own observations and proofs as no more than fallacies and false arguments, is to command something quite impossible for them to do. For it amounts to telling them not to see what they see, and not to understand what they understand and, indeed, to find in their research the very opposite of what evidence shows them. If they were to be asked to do this, they would first have to be shown how to make one mental faculty give orders to another, and the lower faculties to command the higher, so that the imagination and the will were made able and willing to believe the opposite of what the intellect understands (in saying this I am still confining myself to purely scientific questions which are not articles of faith, and not to those which are supernatural and articles of faith). So I do beg these most prudent Fathers to consider very carefully the difference between statements which are a matter of opinion and those which can be demonstrated. If they keep in mind the strength of logical deduction, they will better understand why it is not in the power of those who profess the demonstrative sciences to change their opinion at will, applying themselves first to one view then to another, and that there is a great difference between commanding a mathematician or a philosopher and persuading a merchant or a lawyer to change their mind. It is not as easy to change one's view of conclusions which have been demonstrated in the natural world or in the heavens, as it is to change one's opinion on what is or is not permissible in a contract, a declaration of income, or a bill of exchange. The Church Fathers understood this very well, as can be seen from the great care they took to refute many arguments, or rather fallacies in philosophy. This may be found expressly in some of them; in particular, we have the following words of St Augustine:

> It is unquestionable that whatever the sages of this world have demonstrated concerning physical matters, we can show not to be contrary to our Scripture. But whatever they teach in their books that is contrary to Holy Scripture is without doubt wrong and, to the best of our ability, we should make this evident. And let us keep faith in our Lord, in whom are hidden all the treasures of wisdom, so that we will not be led astray by the glib talk of false philosophy or frightened by the superstition of counterfeit religion.

From these words, it seems to me, the following principle can be derived: that the writings of secular scholars contain some statements about the natural world which are demonstrably true, and others which are simply asserted. As regards the former, it should be the task of wise theologians to show that they are not contrary to Scripture; and as regards the latter—those which are stated but not conclusively demonstrated—if there is anything in them which is contrary to Scripture, they should be regarded as undoubtedly false, and their falseness should be demonstrated by all possible means. Now if scientific conclusions which are demonstrated to be true should not be made subordinate to Scripture, but rather the text of Scripture should be shown not to be contrary to such conclusions, it follows that before a scientific statement is condemned it must be shown that it has not been conclusively demonstrated. And the responsibility for showing this must lie not with those who uphold its truth but with those who believe it to be false: this is only reasonable and natural, for it is much easier for those who do not believe a statement to identify its weaknesses than for those who believe it to be true and conclusive. Indeed, the upholders of an opinion will find that the more they go over the arguments, examining their logic, replicating their observations, and comparing their experiments, the more they will be confirmed in their belief. Your Highness knows what happened when the former mathematician at the University of Pisa undertook in his old age to examine the teaching of Copernicus, in the hope of finding grounds for refuting it (for he was secure in his conviction that it was false as long as he had not read it): as soon as he grasped the foundations, the logic, and the demonstrations of the argument he became convinced by it, and from being an opponent of Copernicus' theory he became its staunch supporter. I could also name other mathematicians who, prompted by my latest discoveries, have acknowledged it necessary to change the accepted system of the world, as it was now completely unsustainable.

If all that was needed to suppress this theory and its teaching was simply to gag a single author, as seems to be the impression of those who, measuring other people's judgement by the standards of their own, cannot believe that it could continue to find supporters, this would be easily done. But the reality is quite different. To achieve such an effect it would be necessary not just to ban Copernicus' book and those of the other authors who have followed his teaching, but to forbid the whole science of astronomy itself. More than that, they would have to forbid men to look at the sky, lest they should see Mars and Venus varying so much in their distance from the Earth that Venus appears forty times, and Mars sixty times, larger at some times than at others. Or they would have to prevent them from seeing Venus appear sometimes round and sometimes crescent-shaped with very fine horns, and many other observations of the senses which are completely incompatible with the Ptolemaic system, but provide solid evidence for the Copernican one. To ban Copernicus' book now, when many new observations and the work of many scholars who have read it are establishing the truth of his position and the soundness of his teaching more firmly every day, and after allowing it to circulate freely for many years when it had few followers and less evidence to support it, would in my view seem to be a contravention of the truth. It would be trying all the harder to conceal and suppress it the more it is plainly and clearly demonstrated. Not to ban the whole book, but just to condemn this particular proposition as false, would, if I am not mistaken, be even more harmful to people's souls, for it would allow them to see the proof of a proposition which they were then told it was sinful to believe. And to forbid the whole science of astronomy would be nothing less than contradicting a hundred passages of Holy Scripture, which teach us that the glory and greatness of God is wonderfully revealed in all his works, and made known divinely in the open book of the heavens. Nor should anyone think that the lofty concepts which are to be found there end in simply seeing the splendour of the Sun and the stars in their rising and setting, which is as far as the eyes of brutes and the common people can see. The book of the heavens contains such profound mysteries and such sublime concepts that all the burning of midnight oil, all the labours, and all the studies undertaken by hundreds of the most acute minds have still not fully penetrated them, even after investigations which have continued for thousands of years. So let even the ignorant recognize that, just as what their eyes see when they look at the external appearance of the human body is as nothing compared to the marvellous complexity which is apparent to the trained and dedicated anatomist and philosopher, who never ceases to be amazed and delighted as he investigates the uses of the muscles, tendons, nerves, and bones, or when he examines the functions of the heart and the other principal organs, seeking out the seat of the vital faculties, observing the wonderful structure of the sensory organs, and contemplating where the imagination, the memory and the power of reason dwell—in the same way, the heavens as they appear to the naked eye are as nothing compared to the great wonders which, through long and painstaking observations, the minds of intelligent men can discern there. This concludes what I have to say on this point.

Next let us answer those who assert that those scientific propositions of which Scripture consistently says the same thing, and which the Church Fathers have all received in the same way, should be understood according to the bare meaning of the words, without trying to gloss or interpret them, and should be accepted and believed as absolutely true; and that the motion of the Sun and the fixity of the Earth are such propositions and are therefore to be believed as matters of faith, and the contrary opinion is to be considered an error. On this I would make the following observations. First, there are some scientific propositions about which human speculation and reason cannot arrive at securely demonstrated knowledge, but can only supply a probable opinion and a reasonable conjecture—such as, for example, whether the stars are animate beings. There are other propositions of which we have, or can confidently expect to have certain knowledge, by means of experiment, prolonged observation, and necessary demonstrations; such are the questions whether the Earth or the Sun moves, or whether the Earth is a sphere. As far as the first kind of proposition is concerned, I have no doubt that where human reason cannot reach, and where consequently we cannot have certain knowledge, but only an opinion or belief, we ought reverently to submit to the pure meaning of Scripture. But as regards the others, I believe that, as I have said above, we must first be certain of the facts, which will reveal to us the true meaning of the Scriptures, which will undoubtedly prove to be in agreement with the demonstrated facts, even if the surface meaning of the words appears to suggest otherwise; for two truths can never contradict each other. This principle seems to me all the more sound and secure because I find it stated in as many words by St Augustine. Writing specifically about the shape of the heavens and what should be believed about it, since astronomers who say that it is round appear to contradict Scripture which states that the heavens are stretched out like a skin, he says that there is no reason to be concerned if Scripture contradicts the astronomers. The authority of Scripture is to be believed if what they say is false and founded only on fallible human conjecture; but if what they affirm is proved by incontrovertible arguments, he does not say that the astronomers are to be ordered to undermine their own proofs and declare their conclusions to be false. Rather, he says that when Scripture describes the heavens as being like a skin it must be shown that this is not contrary to what the astronomers have demonstrated to be true. These are his words:

> But someone may ask: "Is not Scripture opposed to those who hold that heaven is spherical, when it says, 'who stretches out the heavens like a skin'?" It does oppose Scripture if their statement is false, for the truth is rather in what God reveals than in what groping men surmise. But if they are able to establish their position with proofs that cannot be denied, we must show that what is said about the skin is not opposed to the truth of their conclusions.

He goes on to warn us that we should be no less careful to harmonize a passage of Scripture with a demonstrated scientific truth, as with another passage of Scripture which appears to state the opposite. Indeed, I think we should admire the prudence with which this saint, even when he is dealing with difficult questions about which we may be sure that no certain knowledge can be arrived at by human proof, is very cautious about laying down what should be believed. This is what he says at the end of the second book *On the Literal Meaning of Genesis*, on the question of whether we should believe the stars to be animate:

> Although this problem at present is not easy to solve, yet I believe that in the course of our study of Scripture we may come across relevant passages that will enable us to treat the matter according to the rules for interpreting Holy Scripture and arrive at some conclusion that may be held without perhaps demonstrating it as certain. Meanwhile we should always observe that restraint that is proper to a devout and serious person and not rashly believe something about an obscure point. Otherwise, if the truth later becomes known we might despise it because of our attachment to our error, even if what is said is in no way opposed to the sacred writings of the Old or the New Testament.

From this and from other passages it seems to me, if I am not mistaken, that the view of the Church Fathers was that on questions of natural science which are not matters of faith, we should first consider whether they have been demonstrated beyond doubt or are known from the evidence of the senses, or whether such certain knowledge is possible. If it is, and since this too is a gift of God, we should apply ourselves to understanding the true meaning of Scripture in those places where it appears to state the opposite. Wise theologians will undoubtedly be able to penetrate its true meaning, together with the reasons why the Holy Spirit should sometimes have chosen to veil it under words signifying something different, either to test us or for some other reason which is hidden from me.

As for the point about Scripture consistently saying the same thing, I do not think that this should undermine this principle, if we consider the primary intention of Scripture. If it was necessary on one occasion for Scripture to pronounce on a proposition with words conveying a different sense from its true meaning, as a concession to the understanding of the masses, then why might it not have done the same, and for the same reason, whenever the same proposition was mentioned? Indeed, to do otherwise would only have added to people's confusion and undermined their readiness to believe. Regarding the state of rest or motion of the Sun and the Earth, experience plainly shows that it was necessary for Scripture to state what its words appear to say; for even in our own time, people far less primitive still maintain the same opinion, for reasons which on careful consideration and reflection will be found to be wholly trivial, and on experiences which are either erroneous or completely irrelevant. And there is no point in even trying to persuade them to change their view, since they are not capable of understanding the arguments against it, depending as these do on observations which are too precise, proofs which are too subtle, and abstractions which require too much power of imagination for them to comprehend. So even if the fixity of the Sun and the motion of the Earth were established and demonstrated with absolute certainty among the wise, it would still be necessary to uphold the opposite to maintain one's credibility among the vast number of the masses. For if you were to quiz a thousand men among the common people about their view of this matter, I doubt whether you would find one who did not declare himself firmly convinced that the Sun moves and the Earth stands still. But no one should take this almost universal popular consent as an argument for the truth of what they assert; for if we were to question these same men about their grounds and reasons for believing as they do, and on the other hand to listen to the experiments and proofs which have led a few others to believe the opposite, we would find that the latter are persuaded by solidly based reasons, while the former are influenced by shallow appearances and vain and ridiculous comparisons.

It is clear, then, that it was necessary to attribute movement to the Sun and rest to the Earth so as not to confuse the limited understanding of the masses, making them stubborn and reluctant to believe in the principal articles which are absolutely matters of faith; and if this was necessary, then it is not surprising that it was done, with great prudence, in Holy Scripture. I would go further, and say that it was not only consideration for the incomprehension of the masses but the prevailing opinion at that time which led the scriptural writers to accommodate themselves, in matters not necessary to salvation, more to received opinion than to the essential truth of the matter. Speaking of this, St Jerome writes, "as if there were not many things in the Holy Scriptures that were said according to the opinion of the time when they took place, rather than according to the truth contained"; and elsewhere, "It is the practice in the Scriptures for the writer to give the view of things as they were universally believed at that time." And St Thomas, commenting on the words in Job chapter 27,[2] "He stretches out the north over the void, and hangs the earth upon nothing," notes that Scripture refers to the space which enfolds and surrounds the Earth as "void" or "nothing," while we know that it is not empty but full of air. Nonetheless, he says, Scripture adapts to the view of the masses, who believe this space to be empty, by calling it "void" and "nothing." In St Thomas's own words, "What seems to us in the upper hemisphere of the sky to be nothing but space filled with air, the common people consider to be empty; and Holy Scripture speaks of it according to the belief of the common people, as is its wont." From this example I think it can clearly be deduced that on the same principle, holy Scripture had all the more reason to refer to the Sun as in motion and the Earth at rest; for if we challenge the common people's understanding, we will find them much more resistant to the idea that the Sun is at rest and the Earth in motion than to the space around us being full of air. So, if the Scriptural authors refrained from trying to convince the common people even of a point about which they could be persuaded relatively easily, it seems only reasonable that they should have followed the same policy in other much more difficult questions.

Copernicus himself recognized how much our imagination is influenced by ingrained habit and by ways of conceiving things which have been familiar to us since childhood; so in order not to make these abstract ideas even more confusing and difficult for us, once he had demonstrated that the movements which appear to us to belong to the Sun and the firmament are actually movements of the Earth, he continued to call them movements of the Sun and the heavens when he came to set them down in tables and show how they worked in practice. So he talks about the rising and setting of the Sun and the stars, of changes in the inclination of the zodiac and variations in the equinoctial points, of the mean motion, anomalies, and prosthaphaeresis of the Sun, and so on. All these are in fact movements of the Earth; but since we are on the Earth and hence share in its every motion, we cannot discern them in the Earth directly, but have to refer them to the heavenly bodies where they appear to be. Hence we speak of them as if they occurred where we perceive them to

---

2. Galileo cites the wrong chapter here. The quotation actually comes from Job 26:7.

be. This shows how natural it is to adapt ourselves to our habitual way of seeing things.

As for saying that when the Church Fathers agree in interpreting a statement in Scripture on a matter of natural science in the same way this should give it such authority that it becomes a matter of faith to believe it, I think that this should apply at most to those conclusions which the Fathers have aired and discussed exhaustively, weighing up the arguments on both sides before all agreeing that one view should be upheld and the other condemned. But the motion of the Earth and the fixity of the Sun are not propositions of this kind, for such an opinion was completely buried and far removed from the questions discussed by scholars at that time, and was not even considered, let alone upheld, by anyone. So it is fair to assume that it never occurred to the Fathers to discuss the matter, since Scripture, their own views, and the common consent of everyone all agreed on the same opinion, without anyone thinking to contradict it. So it is not enough to say that because the Fathers all accept the fixity of the Earth, etc., this is to be believed as an article of faith; it must be proved that they condemned the contrary opinion. For I could always say that as they never had any occasion to reflect on the matter or to discuss it, they simply left it and accepted it as the current opinion, not as something which had been resolved and established. Indeed, I think I have firm grounds for saying this; for either the Fathers reflected on this as a matter of controversy, or they did not. If they did not, then they cannot have reached any judgement about it, even in their own minds, and their indifference to it should not place any obligation on us to accept precepts which they did not even consider imposing. If on the other hand they had turned their minds to it and considered it, they would already have condemned it if they had judged it to be erroneous, and there is no record of their having done so. Indeed, once some theologians began to consider the matter, it is clear that they did not deem it to be erroneous: Didachus of Stunica, for instance, in his *Commentaries on Job*, chapter 9, verse 6, commenting on the words "Who shakes the earth out of her place," etc., discusses the Copernican position at length and concludes that the motion of the Earth is not contrary to Scripture.

I have, in any case, some reservations about the truth of the claim that the Church requires us to believe as articles of faith such conclusions in natural science as are supported solely by the common interpretation of the Church Fathers. I wonder whether those who argue in this way may have been tempted to extend the scope of the Conciliar decrees in support of their own opinion; for the only prohibition I can find on this matter is against distorting in a sense contrary to the teaching of the Church and the common consent of the Fathers those passages, and those alone, which concern matters of faith or morals or the building up of Christian doctrine. This is what was stated by the Council of Trent in its fourth Session. But the motion or fixity of the Earth or the Sun are not matters of faith or morals, and no one is trying to distort the meaning of Scripture in ways contrary to the teaching of the Church or the Fathers. In fact those who have written about this matter have never cited passages of Scripture, leaving it to the authority of wise and learned theologians to interpret these passages according to their true meaning. It is clear that the Conciliar decrees agree with the Church Fathers in this respect; indeed, so far are they from making such scientific questions articles of faith and condemning contrary opinions as erroneous that they consider it pointless to try to arrive at certainty in such matters, preferring rather to concern themselves with the primary intention of the Church. Let your Highness hear what St Augustine says in response to those Christians who ask whether it is true that the heavens move or whether they are at rest:

> My reply is that it would require a great deal of subtle and learned enquiry into these questions to arrive at a true view of the matter. I do not have the time to go into these questions and nor have those whom I wish to instruct for their own salvation and for what is necessary and useful in the Church.

But even if it were resolved to condemn or admit propositions in natural science according to passages of Scripture which have been unanimously interpreted in the same way by all the Church Fathers, I do not see that this would apply to the present case, for the Fathers differ in their interpretation of the same passages. Dionysius the Areopagite says that it was not the Sun but the Primum Mobile which stood still. St Augustine is of the same opinion, namely that all the celestial bodies came to a stop; so too is the Bishop of Avila. But there are Jewish writers, cited with approval by Josephus, who maintain that the

Sun did not really stand still, but only seemed to because the Israelites took so little time to defeat their enemies. Similarly with the miracle at the time of Hezekiah, Paul of Burgos says that it was not the Sun that moved but the sundial. But in any case, I will show below that it is necessary to gloss and interpret the meaning of the text of the book of Joshua regardless of the view we take of the structure of the universe.

But let us finally concede to these gentlemen more than they ask, and submit entirely to the judgement of wise theologians; and since there is no record of this particular debate being conducted by the ancient Fathers, let it be undertaken by the wise men of our own age. After hearing the experiences, the observations, the arguments, and the proofs cited by philosophers and astronomers on both sides—for it is a controversy over problems of natural science and logical dilemmas, in which a decision has to be made one way or the other—they will be able to determine the matter positively as divine inspiration dictates. But as for those who are ready to risk the majesty and dignity of Holy Scripture for the sake of defending their own vain imagination, let them not hope that such a resolution as this is to be reached without establishing the facts with certainty and discussing in detail all the reasons on both sides of the argument; nor need those who seek only that the foundations of this teaching should be carefully considered, prompted purely by a holy zeal for the truth, for Scripture, and for the majesty, dignity, and authority in which all Christians are bound to uphold it, have anything to fear from such a procedure. Surely it is plain to see that this dignity is far more zealously sought and secured by those who submit whole-heartedly to the Church, without asking for one or other opinion to be prohibited but only that they should be allowed to bring matters forward for discussion so that the Church can reach a decision with greater confidence, than by those who, blinded by their own self-interest or prompted by the malicious suggestions of others, preach that the Church should wield its sword straight away simply because it has the power to do so? Do they not realize that it is not always beneficial to do what one has power to do? This was not the view of the Church Fathers; on the contrary, they knew how prejudicial and how contrary to the primary intention of the Catholic Church it would be to use verses of Scripture to establish scientific conclusions which experience and necessary demonstrations might in time show to be contrary to the literal meaning of the text. Hence they not only proceeded with great caution, but they also left the following precepts for the guidance of others:

> In matters that are obscure or far from clear, if we should read anything in Holy Scripture that may allow of different interpretations that are consistent with the faith we have received, we should not rush in headlong and so firmly commit ourselves to one of these that, if further progress in the search of truth justly undermines this position, we too fall with it. That would be to battle not for the meaning of Holy Scripture but for our own, by wanting something of ours to be the meaning of Scripture rather than wanting the meaning of Scripture to be ours.

A little further on, to teach us that no proposition can be contrary to the faith if it has not first been shown to be false, he adds: "Nothing is contrary to the faith until unerring truth gives the lie to it. And if that should happen, it was never taught by Holy Scripture but stemmed from human ignorance." It is clear from this that any view which we attributed to a passage of Scripture would be false if it did not agree with demonstrated truth. Therefore we should use demonstrated truth to help us discover the correct meaning of Scripture, and not try to force nature or deny the evidence of experience and necessary demonstrations in order to conform to the literal meaning of the words, which our imperfect understanding might think to be true.

But note further, your Highness, how carefully this great saint proceeds before affirming that a particular interpretation of Scripture is correct, and so firmly established that there need be no fear of encountering any difficulty which might undermine it. Not content that a reading of Scripture should agree with a demonstrated truth, he adds:

> But when some truth is demonstrated to be certain by reason, it will still be uncertain whether this sense was intended by the sacred writer when he used the words of Holy Scripture, or something else no less true. And if the general drift of the passage shows that the sacred writer did not intend this sense, the other, which he did intend, will not thereby be false. Indeed, it will be true and more worth knowing.

Yet this author's caution is even more remarkable when, not being convinced after seeing the demonstrations, the literal meaning of Scripture, and the context of the passage as a whole all pointing to the same interpretation, he adds: "But if the context supplies nothing to disprove this to be the mind of the writer, we still have to enquire whether he may not have meant the other as well." Not even then resolving to accept one interpretation and reject the other, he seems to think he can never be cautious enough, for he goes on: "But if we find that the other also may be meant, it will not be clear which of the two meanings he intended. And there is no difficulty if he is thought to have wished both interpretations if both are supported by clear indications in the context." Finally, he justifies this rule of his by showing the dangers to which Scripture and the Church are exposed by those who, being more interested in maintaining their own error than in upholding the dignity of Scripture, seek to extend the authority of Scripture beyond the terms which Scripture itself prescribes. He adds the following words, which alone should suffice to restrain and moderate the excessive licence which some claim for themselves:

> It often happens that a non-Christian knows something about the earth, the heavens, and the other elements of this world, about the motion and orbit of the stars and even their size and relative positions, about the predictable eclipses of the Sun and Moon, the cycles of the years and the seasons, about the kinds of animals, shrubs, stones, and so forth, and this knowledge he holds to as being certain from reason and experience. Now, it is a disgraceful and dangerous thing for an infidel to hear a Christian, presumably giving the meaning of Holy Scripture, talking nonsense on these topics. We should take all means to prevent such an embarrassing situation in which the non-believer will scarce be able to contain his laughter seeing error written in the sky, as the proverb says. The shame is not so much that an ignorant individual is derided, but that people outside the household of the faith think our writers hold such opinions, and criticize and reject them as ignorant, to the great prejudice of those whose salvation we are seeking. When they find a Christian mistaken in a field which they themselves know well and hear him maintaining foolish opinions about our books,

how are they going to believe those books in matters concerning the resurrection of the dead, the hope of eternal life, and the kingdom of heaven, when they think their pages are full of falsehoods about things which they themselves have learnt from experience and decisive argument?

This same saint shows how much the truly wise and prudent Fathers are offended by those who try to uphold propositions which they do not understand by citing passages of Scripture, compounding their original error by producing other passages which they understand even less than the first; he writes:

> Rash and presumptuous men bring untold trouble and sorrow on their wiser brethren when they are caught in one of their false and unfounded opinions and are taken to task by those who are not bound by the authority of our sacred books. For then, to defend their utterly reckless and obviously untrue statements, they call upon Holy Scripture, and even recite from memory passages which they think support their position, although they understand neither what they mean nor to what they properly apply.

This seems to me to describe exactly those who keep citing passages of Scripture because they are unable or unwilling to understand the proofs and experiments which the author of this doctrine and his followers advance in its support. They do not realize that the more passages they cite and the more they insist that their meaning is perfectly clear and cannot possibly admit any other interpretation than theirs, the more they would undermine the dignity of Scripture (if, that is, their opinions carried any weight) if the truth were then clearly shown to contradict what they say, causing confusion at least among those who are separated from the Church and whom the Church, like a devoted mother, longs to bring back to her bosom. So your Highness can see how flawed is the procedure of those who, in debating questions of natural science, give priority in support of their arguments to passages of Scripture—and often passages which they have misunderstood.

But if they really believe and are quite certain that they possess the true meaning of a particular text of Scripture, they must necessarily be convinced that they hold in their hand the absolute truth of the scientific conclusion which they intend to debate, and so must know that they have a great advantage over their opponent who has to defend what is false. The one who is defending the truth will be able to draw on numerous sensory experiences and necessary demonstrations to support their position, while their opponent has to fall back on deceptive appearances, illogical reasoning, and fallacies. Why then, if they are so confident that their purely scientific and philosophical weapons are so much stronger in every way than their adversary's, do they immediately have recourse, as soon as battle is joined, to an awesome and irresistible weapon the very sight of which strikes terror into the heart of their opponent? If the truth be told, I believe that they are the first to be terror-struck, and that when they realize they are unable to resist the assaults of their adversary they try to find a way of not letting him come near them. To that end they forbid him to use the gift of reason which the divine goodness has granted him, and abuse the right and proper authority of Scripture which by common consent of theologians, if it is understood and used properly, can never contradict the evidence of plain experience and necessary demonstrations. But I do not think that their resorting to Scripture to cover up their inability to understand, let alone to answer, the arguments against them, will do them any good, since this opinion has never hitherto been condemned by the Church. So if they wish to deal honestly they should either confess by their silence that they are unqualified to discuss such matters, or they should first consider that it is not in their power or that of anyone except the Supreme Pontiff or the Councils of the Church to declare a proposition to be erroneous, although they do have the right to debate whether it is true or false. Then, since it is impossible for any proposition to be both true and heretical, they should concern themselves with what they are entitled to discuss, namely demonstrating that it is false. Once they have established its falsehood, either there will be no more need to prohibit it because no one will subscribe to it, or it can safely be prohibited without any risk of causing scandal.

So let these people first apply themselves to refuting the arguments of Copernicus and others, and leave condemning his view as erroneous and heretical to those who have the authority to do so; but let them not hope to find in the wise and cautious Fathers of the Church or in the absolute wisdom of the One who cannot err those hasty judgements into which they are sometimes drawn by their own desires or vested interests. No one doubts that the Supreme Pontiff always has absolute power to admit or to condemn these and similar propositions which are not directly articles of faith; but it is beyond the power of any created being to make them true or false, in defiance of what they are de facto by their own nature. So they would be better advised first to establish with certainty the necessary and immutable truth of the matter, over which no one has any control, than to condemn either side in the absence of any such certainty. This would only deprive them of their own authority and freedom to choose, by imposing necessity on matters which at present are undetermined and a subject of free choice but still reserved to the authority of the Supreme Pontiff. In short, if it is not possible for a conclusion to be declared heretical while there is still uncertainty over whether it may be true, it is a waste of time for anyone to clamour for the condemnation of the motion of the Earth and the fixity of the Sun before they have demonstrated that such a position is impossible and false.

It remains finally for us to consider how far the passage in Joshua can be taken in the straightforward literal meaning of the words, and how it could come about that the day was much prolonged when the Sun obeyed Joshua's command to stand still.

If the movements of the heavens are taken according to the Ptolemaic system, such a thing cannot happen. For the Sun moves through the ecliptic in the order of the signs of the zodiac, that is, from west to east, and hence in the opposite direction to the motion of the Primum Mobile, which is from east to west, this being the motion which produces day and night. It is clear therefore that if the Sun were to cease its own proper motion, the effect would be to make the day shorter, not longer. The way to make the day longer would be to speed up the Sun's motion; for to make the Sun remain at the same point above the horizon for some time without declining towards the west, its motion would have to be speeded up until it equalled that of the Primum Mobile, which would be about 360 times its normal speed. So if Joshua had meant his words to be taken in their strict literal sense, he would have commanded the Sun to speed up its motion so that the Primum Mobile ceased to carry it along towards its setting. But since his words were heard by people whose knowledge of the motions of the heavens was very likely confined to just the universally known movement from east to west, and since he had no intention of teaching them about the structure of the spheres but only that they should comprehend the great miracle of prolonging the day, he adapted his words to their understanding and spoke in the way which would make sense to them.

Perhaps it was this consideration which prompted Dionysius the Areopagite to say that the miracle consisted in stopping the Primum Mobile, with the consequence that all the other celestial spheres also stood still; St Augustine himself is of the same opinion, and the Bishop of Avila also confirms it at length. Indeed it is clear that Joshua's intention was to make the whole system of celestial spheres stand still, because his command also included the Moon even though the Moon had nothing to do with prolonging the day. By commanding the Moon he implicitly included all the other planets, which are not named here any more than they are elsewhere in Scripture, since it has never been the intention of Scripture to teach us the science of astronomy.

It seems clear to me, therefore, if I am not mistaken, that if we were to accept the Ptolemaic system, we would have to interpret the words of Scripture in a sense different from their literal meaning; but bearing in mind the salutary warnings of St Augustine, I do not say that this is necessarily the correct interpretation, as someone else might come up with a better and more appropriate one. I would, however, like to conclude by asking whether this passage can be understood in a sense closer to what we read in Joshua if we assume the Copernican system, together with a further observation which I have recently made concerning the body of the Sun. I put forward this suggestion always with the reservation that I am not so wedded to my own ideas as to claim they are superior to other people's, or to deny that better interpretations may be forthcoming which would conform more closely to the intention of Holy Scripture.

Let us assume, then, in the first place, that the miracle in the book of Joshua meant bringing the whole system of celestial revolutions to a standstill, as suggested by the authors quoted above; this is because if just one sphere were to stand still it would upset the whole system, introducing unnecessary disruption throughout the whole of nature. Secondly, I take into account that the body of the Sun, while remaining fixed in the same place, nonetheless turns on its own axis, completing one revolution in about a month, as I believe I have demonstrated in my *Letters on the Sunspots*. We can observe this movement and see that in the upper part of the Sun's globe it is inclined towards the south, and therefore in the lower part it inclines towards the north—in just the same manner as all the revolutions of the planets. Thirdly, if we consider the nobility of the Sun and the fact that it is the source of light, not only for the Moon and the Earth but, as I show conclusively, also for all the other planets, all of which similarly have no light of their own, I think it is not unreasonable to suggest that the Sun, as the chief minister of nature and, in a sense, the heart and soul of the universe, dispenses not only light to the bodies which surround it but also motion, by virtue of its turning on its own axis. This means that, just as if an animal's heart stopped beating all the other parts of the body would also stop moving, so if the rotation of the Sun were to cease, the rotations of the planets would stop as well. Of the many weighty writers I could cite to confirm the wonderful strength and power of the Sun, let it suffice for me to quote one passage from the blessed Dionysius the Areopagite in his book *On the Divine Names*. Writing of the Sun, he says: "His light also gathers and converts to itself all the things that are seen, moved, lighted or heated, in a word everything that is held together by its splendour. Therefore the Sun is called Helios, for it gathers and brings together everything that is scattered." A little later he goes on to say,

> This Sun which we see is one, and although the essences and qualities of those things that we perceive with our senses are many and varied, yet the Sun sheds its light equally on all things, and renews, feeds, protects, completes, divides, unites, fosters, makes fruitful, increases, changes, fixes, builds up, moves, and gives life to them all. Every single thing in this universe, inasmuch as it can, partakes of one and the same Sun, and the causes of many things which partake of it are equally anticipated in it; and for all the more reason, etc.

So, the Sun being the source of both light and motion, if God wished that at Joshua's command the whole system of the world should rest and remain for several hours in the same state, it sufficed to make the Sun stand still. When the Sun stopped, all the other revolutions stopped as well; the Earth, Moon and Sun remained in the same relationship to each other, as did all the other planets; and as long as this continued the day did not decline towards night, but was miraculously prolonged. In this way, by stopping the Sun it was possible to prolong the day on Earth, without altering or disrupting the other aspects and mutual positions of the stars, which agrees perfectly with the literal sense of the sacred text.

Another point which, if I am not mistaken, is of no small significance, is that the Copernican system makes another detail in the literal account of this miracle perfectly clear: namely, that the Sun stood still "in the midst of the heavens." This passage has caused learned theologians some difficulty, because it seems likely that when Joshua prayed for the day to be prolonged it was already near to sunset, not at midday—for if it had been at midday, and given that this happened about the time of the summer solstice when the days are at their longest, it seems unlikely that he would have needed to pray for the day to be prolonged so that he could pursue the battle to victory; the seven hours or more of daylight which remained would have been more than enough. So some very learned theologians have concluded that it must have been near to sunset; and indeed this is implied by Joshua's words, "Sun, stand still," since if it had been at midday, either he would not have needed to ask for a miracle or it would have been enough to pray for the Sun to slow down. This is the view of Cajetan, to which Magalhães also subscribes, and confirms it by pointing out that Joshua had already done so many other things that day before he commanded the Sun that he could not possibly have completed them all in half a day. So they are reduced to interpreting the words "in the midst of heaven" in a somewhat forced way, saying that they simply mean the Sun stopped when it was in our hemisphere, that is, when it was above the horizon. But I think we can avoid this and any other forced reading if we follow the Copernican system and place the Sun "in the midst," that is, in the centre of the heavenly orbs and the revolutions of the planets, as indeed we must. Then, regardless of the time of day, whether at midday or at any other time towards evening, the day was prolonged and the revolutions of the heavens stood still when the Sun stopped in the midst of heaven, that is, in the centre, where it belongs. Apart from anything else, this is a more natural reading of the literal sense of the text, for if the writer had wanted to say that the Sun stood still at noon it would have been more correct to say that it "stood still at midday, or in the circle of the meridian," not "in the midst of heaven." For the only true "midst" of a spherical body like the sky is its centre.

As for other passages of Scripture which appear to contradict the Copernican position, I have no doubt that, if this position were once known to be true and proven, those same theologians who now, believing it to be false, find such passages incapable of being interpreted in a way compatible with it, would find interpretations for them which would accord with it very well, especially if their understanding of Holy Scripture were combined with some knowledge of astronomy. Just as now, believing this position to be false, they read the Scriptures and find only passages which conflict with it, so if they once entertained a different view of the matter they might well find just as many others which agreed with it. Then they might judge it fitting for the holy Church to proclaim that God placed the Sun in the centre of the heaven and, by turning it on its axis like a wheel, gave the Moon and the other wandering stars their appointed course, when she sings the hymn:

> O God, whose hand hath spread the sky,
> and all its shining hosts on high,
> and painting it with fiery light,
> made it so beauteous and so bright:
>
> Thou, when the fourth day was begun,
> didst frame the circle of the sun,
> and set the moon for ordered change,
> and planets for their wider range.

They could also say that the word "firmament" is literally correct for the starry sphere and for everything which is beyond the revolutions of the planets, for in the Copernican system this is totally firm and immobile. And since the Earth moves in a circle, when they read the verse, "Before he had made the Earth and the rivers, and the hinges of the earth," they might think of its poles, for it seems pointless to attribute hinges to the terrestrial globe if it does not turn on its axis.

Bibliography and Additional Readings

Anastos, Milton V. "The History of Byzantine Science: Report on the Dumbarton Oaks Symposium of 1961." *Dumbarton Oaks Papers* 16 (1962): 409–11. http://www.jstor.org/stable/1291170?seq=1#page_scan_tab_contents.

Atkins, Peter W. "The Limitless Power of Science." In *Nature's Imagination: The Frontiers of Scientific Vision*, edited by John Cornwell, 122–32. New York: Oxford University Press, 1995.

Block, David L. "Georges Lemaître and Stigler's Law of Eponymy." In *Georges Lemaître: Life, Science and Legacy*, edited by Rodney D. Holder and Simon Mitton, 89–96. New York: Springer, 2012.

———. *Our Universe: Accident or Design?* Edinburgh: Scottish Academic Press, 1992.

———. "Rings in Spiral Galaxies in the Local Group: Lessons from René Magritte." In *Lessons from the Local Group: A Conference in Honour of David Block and Bruce Elmegreen*, edited by Kenneth C. Freeman, Bruce G. Elmegreen, David L. Block, and Matthew Woolway, 423–41. New York: Springer, 2015.

Block, David L., and Kenneth Freeman. *Shrouds of the Night: Masks of the Milky Way and Our Awesome New View of Galaxies*. New York: Springer, 2008.

———. "A Walk with Dr Allan Sandage: Changing the History of Galaxy Morphology, Forever." In *Lessons from the Local Group: A Conference in Honour of David Block and Bruce Elmegreen*, edited by Kenneth C. Freeman, Bruce G. Elmegreen, David L. Block, and Matthew Woolway, 1–20. New York: Springer, 2015.

Bontrager, Scot C. "Nature and Grace in the First Question of the *Summa*." *Scot Bontrager* (blog), February 1, 2010. https://www.indievisible.org/Papers/Aquinas%20-%20Nature%20and%20Grace.pdf.

Bucciantini, Massimo, Michele Camerota, and Franco Giudice. *Galileo's Telescope: A European Story*. Translated by Catherine Bolton. Cambridge, MA: Harvard University Press, 2015.

Campbell, Arabella Georgina. *The Life of Fra Paolo Sarpi*. London: Molini and Green, 1869.

Carr, H. Wildon. "The Tercentenary of Blaise Pascal." *Nature* 111, no. 2798 (1923): 814.

Carroll, John W., ed. *Readings on Laws of Nature*. Pittsburgh: University of Pittsburgh Press, 2004.

Carter, Brandon. "The Anthropic Principle and Its Implications for Biological Evolution." *Philosophical Transactions of the Royal Society of London*. Ser. A, *Mathematical and Physical Sciences* 310, no. 1512 (1983): 347.

Chesterton, G. K. *The Everlasting Man*. New York: Image Books, 1955.

———. *St. Thomas Aquinas: The Dumb Ox*. New York: Sheed and Ward, 1933.

———. "The Wind and the Trees." In *Tremendous Trifles*, 61–65. 1909. Reprint, London: Methuen, 1930.

Cowell, Alan. "After 350 Years, Vatican Says Galileo Was Right: It Moves." *New York Times*, October 31, 1992. http://www.nytimes.com/1992/10/31/world/after-350-years-vatican-says-galileo-was-right-it-moves.html.

Dawkins, Richard. *The Blind Watchmaker*. Harlow, UK: Longman Scientific and Technical, 1986.

Deazley, Ronan, Martin Kretschmer, and Lionel Bently, eds. *Privilege and Property: Essays on the History of Copyright*. Cambridge: Open Book, 2010. http://books.openedition.org/obp/1062?lang=en.

de Fontenelle, Bernard le Bovier. *Conversations with a Lady on the Plurality of Worlds* [*Etretiens sur la Pluralité des Mondes*]. Translated by H. A. Hargreaves. Edited by Nina Rattner Gelbart. Berkeley: University of California Press, 1990. Originally published in French in 1686.

de Santillana, Giorgio. *The Crime of Galileo*. Chicago: University of Chicago Press, 1955.

Disney, Michael J. "The Case against Cosmology." *General Relativity and Gravitation* 32, no. 6 (2000): 1125–34. https://ned.ipac.caltech.edu/level5/Disney/paper.pdf.

Dostoevsky, Fyodor M. *The Grand Inquisitor*. Translated by S. S. Koteliansky. London: Elkin Mathews & Marrot, 1930.

Ellis, George F. R. "Opposing the Multiverse." *Astronomy and Geophysics* 49, no. 2 (2008): 2.5–2.7.

———. "The Thinking Underlying the New 'Scientific' World-Views." In *Evolutionary and Molecular Biology: Scientific Perspectives on Divine Action*, edited by Robert John Russell, William R. Stoeger, and Francisco J. Ayala, 251–80. Vatican: Vatican Observatory Foundation, 1998.

Elmegreen, Bruce. "Observations and Theory of Dynamical Triggers for Star Formation." *Astronomical Society of the Pacific Conference Series* 148 (1998): 150–83.

———. Quoted in "Star Formation in Galaxies." In *The Spectral Energy Distribution of Galaxies: Proceedings of the 284th Symposium of the International Astronomical Union, Held at the University of Central Lancashire, Preston, United Kingdom, September 5–9, 2011*, edited by Richard J. Tuffs and Cristina C. Popescu, 317–29. Cambridge: Cambridge University Press, 2012.

Fantoli, Annibale. *Galileo: For Copernicanism and the Church*. Translated by George V. Coyne. Vatican Observatory Publications. Studi Galleiani 3. Notre Dame, IN: University of Notre Dame Press, 1994.

Finocchiaro, Maurice A. *The Galileo Affair: A Documentary History*. California Studies in the History of Science. Berkeley: University of California Press, 1989.

———, trans. and ed. *The Trial of Galileo: Essential Documents*. Indianapolis: Hackett, 2014.

Foucault, Michel. *This Is Not a Pipe* [*Ceci n'est une pipe*]. Translated and edited by James Harkness. Los Angeles: University of California Press, 1982.

Franklin, Benjamin. "The Morals of Chess." *Columbian Magazine* (1786): 159; online version: McCrary, John. "Chess and Benjamin Franklin: His Pioneering Contributions." http://www.benfranklin300.org/_etc_pdf/Chess_John_McCrary.pdf.

Frost, Robert. "Kitty Hawk." *Atlantic Monthly*, November 1957, 52–56.

Galilei, Galileo. *Galileo: Selected Writings*. Translated by William R. Shea and Mark Davie. Oxford World's Classics. New York: Oxford University Press, 2012.

———. *Istoria e Dimostrazioni Intorno Alle Macchie Solari e Loro Accidenti* [History and demonstrations concerning sunspots and their properties]. 1613. Reprint, Rome: Theoria, 1982.

———. *Le Opere di Galileo Galilei: Edizione Nazionale*. Edited by Antonio Favoro. 1890–1909. Reprint, Florence: Giunti Barbèra, 2013–2015.

———. *Lettera a Cristina di Lorena*. Edizione critica a cura di Ottavio Besomi, collaborazione di Daniele Besomi, versione latina di Elia Diodati a cura di Giancarlo Reggi. Medioevo e umanesimo 116. Rome: Antenore, 2012.

———. *Letter to the Grand Duchess Christina*. In *Discoveries and Opinions of Galileo*, translated and edited by Stillman Drake, 173–216. New York: Anchor Books, 1957. Original Italian text published in *Opere di Galileo*

Galilei: Edizione Nazionale, edited by Antonio Favaro, 5:309–48. Firenze: Giunti-Barbera, 1968.

Gallico, Paul. *Snowflake*. New York: Doubleday, 1953.

Gingerich, Owen. "The Curious Case of the M-L Sidereus Nuncius." *Galilaeana: Journal of Galilean Studies* 6 (2009): 141–65.

———. "The Delights of a Roving Mind." In Johannes Kepler. *The Six-Cornered Snowflake: A New Year's Gift*, edited by Guillermo Bleichmar and Jacques Bromberg, 1–12. Philadelphia: Paul Dry Books, 2010.

———. "The Galileo Affair," *Scientific American* 247, no. 2 (1982): 119–27.

Goodwin, Richard N. *The Hinge of the World: A Drama*. New York: Farrar, Straus and Giroux, 1998.

Haldane, J. B. S. *Possible Worlds and Other Essays*. London: Chatto and Windus, 1927.

Heyl, Norbert, and Rosa Barovier Mentasti. *Murano: The Glass-Making Island*. Translated by Clare Loraine Walford. Ponzano Veneto: Vianello Libri, 2006.

Hooper, Walter. *C. S. Lewis: A Companion and Guide*. London: HarperCollins, 1996.

Keel, William C. *The Road to Galaxy Formation*. 2nd ed. New York: Springer-Praxis, 2007.

Kostylo, Joanna. "From Gunpowder to Print: The Common Origins of Copyright and Patent." In *Privilege and Property: Essays on the History of Copyright*, edited by Ronan Deazley, Martin Kretschmer, and Lionel Bently, chap. 1. Cambridge: Open Book, 2010. http://books.openedition.org/obp/1062?lang=en.

Langford, Jerome J. *Galileo, Science and the Church*. Ann Arbor Paperbacks. Ann Arbor, MI: University of Michigan Press, 1992.

Lewis, C. S. *The Discarded Image: An Introduction to Medieval and Renaissance Literature*. Cambridge: Cambridge University Press, 1964.

———. *Miracles: A Preliminary Study*. London: Geoffrey Bles, 1947.

Libbrecht, Kenneth G. "The Physics of Snow Crystals." *Progress of Reports in Physics* 68, no. 4 (2005): 855–95.

Linder, Douglas O. "Papal Condemnation (Sentence) of Galileo." Famous Trials, accessed September 13, 2018. http://www.famous-trials.com/galileotrial/1012-condemnation.

———. "Trial of Galileo: A Chronology." Famous Trials, accessed June 26, 2018. http://www.famous-trials.com/galileotrial/1015-chronology.

Machamer, Peter. "The Fate of Galileo and His Spyglass." In Galileo Galilei, *The Starry Messenger, Venice 1610: "From Doubt to Astonishment,"* edited by John W. Hessler and Daniel De Simone, 196–205. Delray Beach, FL: Levenger Press in association with the Library of Congress, 2013.

Maffeo, Sabino. *The Vatican Observatory: In the Service of Nine Popes.* Translated by George V. Coyne. 2nd ed. Vatican City: Vatican Observatory Publications, 2001.

McGrath, Alister, and Joanna Collicutt McGrath. *The Dawkins Delusion? Atheist Fundamentalism and the Denial of the Divine.* Downers Grove, IL: IVP Books, 2007.

"The Medici Family." History.com. Updated August 21, 2018. http://www.history.com/topics/medici-family.

Merton, Robert. "Priorities in Scientific Discovery. A Chapter in the Sociology of Science; Presidential Address Read at the American Sociological Society." *Journal of the American Sociological Society* 22, no. 6 (1957): 635–59.

Micanzio, Fulgenzio. *The Life of the Most Learned Father Paul [Sarpi], of the Order of the Servie, Councellour of State to the Most Serene Republicke of Venice.* London: Humphrey Moseley and Richard Marriott, 1651.

Miller, David Marshall. "Seeing and Believing: Galileo, Aristotelians, and the Mountains on the Moon." In Galileo Galilei, *The Starry Messenger, Venice 1610: "From Doubt to Astonishment,"* edited by John W. Hessler and Daniel De Simone, 131–45. Delray Beach, FL: Levenger Press in association with the Library of Congress, 2013.

Milton, John. *Areopagitica.* London: Adam and Charles Black, 1911.

———. *Of Reformation Touching Church-Discipline in England.* 1641. Reprint, New Haven, CT: Yale University Press, 1916.

Morrow, Lance. *Fishing in the Tiber: Essays.* New York: Henry Holt, 1988.

Orr, H. Allen. "Gould on God: Can Religion and Science Be Happily Reconciled?" *Boston Review*, October 1999. http://new.bostonreview.net/BR24.5/orr.html.

Pascal, Blaise. "Pascal's Profession of Faith" [or "Memorial"]. In *The Thoughts of Blaise Pascal*, translated by C. Kegan Paul, 2. London: Kegan Paul, Trench, 1885. http://www.gutenberg.org/files/46921/46921-h/46921-h.htm.

———. *Pensées.* Translated by A. J. Krailsheimer. New York: Penguin, 1966.

Plantinga, Alvin. *Where the Conflict Really Lies: Science, Religion, and Naturalism.* New York: Oxford University Press, 2011.

Pope John Paul II. "Faith Can Never Conflict with Reason." Translated from *L'Osservatore Romano* N. 44 (1264), November 4, 1992. http://www.elabs.com/van/JPII-faith_can_never_conflict_with_reason.htm.

Reeves, Eileen. *Painting the Heavens: Art and Science in the Age of Galileo*. Princeton, NJ: Princeton University Press, 1997.
Rosen, Jonathan. "Return to Paradise: The Enduring Relevance of John Milton." *New Yorker*, June 2, 2008. http://www.newyorker.com/magazine/2008/06/02/return-to-paradise.
Schrödinger, Erwin. *Mind and Matter*. Cambridge: Cambridge University Press, 1958.
Scott, Andrew. *The Creation of Life*. Oxford: Basil Blackwell, 1986.
Stigler, Stephen M. "Stigler's Law of Eponymy." *Transactions of the New York Academy of Sciences*, 2nd ser., 39, no. 1 (1980): 147–57.
Teems, David. *Tyndale: The Man Who Gave God an English Voice*. Nashville: Thomas Nelson, 2012.
Tyndale, William. *The New Testament of Our Lord and Saviour Jesus Christ—Published in 1526—Being the First Translation from the Greek into English, by That Eminent Scholar and Martyr William Tyndale; With a Memoir of His Life and Writings, by George Offor*. London: Samuel Bagster, 1836.
Vulgate. Proverbs 8. Christian Classics Ethereal Library. Accessed October 10, 2018. http://www.ccel.org/ccel/bible/vul.Prov.8.html.
"What Life Means to Einstein: An Interview by George Sylvester Viereck." *Saturday Evening Post*, October 26, 1929, 17, 110, 113–14, 117. http://www.saturdayeveningpost.com/wp-content/uploads/satevepost/einstein.pdf.
Wilding, Nick. *Galileo's Idol: Gianfrancesco Sagredo and the Politics of Knowledge*. Chicago: University of Chicago Press, 2014.
Wootton, David. *Paolo Sarpi: Between Renaissance and Enlightenment*. Cambridge: Cambridge University Press, 1983.
Yates, Frances A. "Paolo Sarpi's *History of the Council of Trent*." *Journal of the Warburg and Courtauld Institutes* 7 (1944): 123–43.

General Index

academics, pride and prejudice of, 175
accommodation, 97–98
"Against Wearing the Gown" (Galileo poem), 167–68, 178
Andromeda galaxy, 102
anthropic principle, 74
Antikythera mechanism, 95
Aquinas, Thomas, 36
Areopagitica (Milton), 81–82
Aristarchus of Samos, 95
Aristotle, 27
armillary sphere, 96
asps, 32
astronomy, 29, 32–33
atheism, under mantle of science, 13, 42, 66, 69
Atkins, Peter, 123, 125
Augustine, 31, 33, 64, 76, 77, 82, 95, 112, 116, 121–22, 126
aurora borealis, 30
authority of the establishment, 33, 35

Baade, Walter, 159–60
Ball, W. W. Rouse, 163
Ballarin, Giorgio, 145
Barnard, Edward Emerson, 173
beliefs, as transitory, 67
Bellarmine, Cardinal, 105n9
Bible. *See also* book of Scripture
　authority of, 61, 69
　as "dumbed down" for the masses, 96–101
　interpretation of, 112
　not a scientific textbook, 28, 35, 47
　physically incomplete allusions to natural phenomena, 96–99
　used to silence science, 13
black holes, 124
Blackwell, Richard, 175
Blake, William, 68

blindness, 41, 42, 43, 78, 81, 82, 169, 178
Block, David, 77, 169–78
blue stars, 124
Bontrager, Scot, 36, 125, 169
book of nature, 27–29, 35, 129, 130, 134–35, 160
　as book of process, 178
　boundaries of, 126
　connection with book of Scripture, 112–14
　not intended to make judgments on book of Scripture, 35
　suppression of, 81
　used to discount book of Scripture, 35
book of Scripture, 27–29, 129, 130, 134–35, 160. *See also* Bible
　as book of purpose, 178
　boundaries of, 126
　connection with book of nature, 112–14
　and salvation, 55
　as source of all truth, 39
Brothers Karamazov (Dostoevsky), 118–19
Brush, Nigel, 29
Bucciantini, Massimo, 91, 139, 142, 143

Caccini, Tommaso, 44, 150
Calvin, John, 153
Camerota, Michele, 91, 139, 142, 143
Campbell, Arabella Georgina, 134
carbon-based life, 40
Carr, H. Wildon, 164
Carter, Brandon, 74
Castel Gandolfo, 156, 159
Castelli, Benedetto, 149, 151
cathedral of science, 58, 62, 129
Catherine de Medici, 25
caution, 116–17
censorship, 80–81
chess, 116, 125
Chesterton, G. K., 36–37, 42, 60–62, 67, 104, 113–14, 131, 132, 136

Christina of Lorraine (Grand Duchess of Tuscany), 25, 149, 152
church
  dispute with Galileo, 23–25
  power of, 53, 96, 99, 127, 130
church fathers, 109
circumspection, 116–17
city of God, 172–73
classical physics, 64
common people, 49–50, 53–55, 97
conflict, between religion and science, 23, 93, 96, 105–7, 113
Copernicanism, 150
Copernicus, Nicolaus, 15, 23, 27, 33–34, 41, 43, 80, 105, 149, 151
Cosimo II de Medici, 25
cosmological principle, 105–6
cosmology, 58
Council of Trent, 109–12
Counter-Reformation, 109–11
Coyne, George, 175, 176

d'Alvise, Bortolo, 145
dark energy, 56, 57, 65
dark matter, 56–57, 59, 65
Dawkins, Richard, 56, 78, 174
de Santillana, Giorgio, 135n19, 160
Descartes, René, 45–46
De Zúñiga, Diego, 105
Dialogue concerning the Two Chief World Systems (Galileo), 153
Diodati, Giovanni, 98, 134
Disney, Michael, 124–25
Dostoevsky, Fyodor, 118–19, 130–31

earth-centered universe, 93–94
Einstein, Albert, 14, 60, 63, 129, 175, 176
Ellis, George, 57, 59, 75
Elmegreen, Bruce, 55, 139, 144
Emerson, Ralph Waldo, 172
Engle, Paul, 147
Erasmus, Desiderius, 50, 53
Esther, 176–77
evolution, 66–67
exclusion principle, 43

Fantoli, Annibale, 149–50, 151–52
figurative speech, 99–100
Findlen, Paula, 140
finely tuned universe, 42, 74
Finocchiaro, Maurice, 52–53, 98, 149, 150, 159
Fontenelle, Bernard le Bovier de, 41

foresight, 116
Foscarini, Antonio, 152
Foucault, Michel, 101–2
Foxe, John, 51
Franklin, Benjamin, 116, 125
Freeman, Kenneth, 178
Frost, Robert, 129, 176
fundamentalism, 128

galaxies, 77, 123–24
Galileo Galilei, 15, 129, 175
  "Against Wearing the Gown," 167–68, 178
  on Augustine, 121
  on authority of Scripture, 61
  on censorship, 80–81
  on church fathers, 126
  on Council of Trent, 109
  death of, 135
  on dignity of theology, 73
  house arrest of, 135
  letter to Castelli, 149–51
  Letter to the Grand Duchess Christina of Tuscany, 15–16, 27, 30, 31, 39, 149, 151–53, 179–213
  on ordinary people, 53–54
  on personal agendas and interests, 129
  and rise of modern science, 72
  Sidereus Nuncius, 54, 141–42, 146
  trial of, 83–91, 158
  on two books, 28, 44, 47
Galileo Commission, 176
Genesis, misinterpretation of, 32
geocentric model of the universe, 23, 27, 33, 112, 120
Gingerich, Owen, 23, 43
Giudice, Franco, 91, 139, 142, 143
glare of science, 123
"Gloria in Profundis" (Chesterton poem), 132
God
  as a delusion, 56, 174
  as "hands-off," 67
  of the macrocosm and microcosm, 131
  not in competition with science, 128
  as virus, 78–79
"God of the gaps," 30, 82, 128
Goodwin, Richard, 133
Gould, Stephen Jay, 113
grace, 14, 119, 120, 128, 164, 171–72, 177
gravity, 40
Gregory XIII, Pope, 155, 156
Grillo, Angelo, 91

harmony, of science and faith, 25, 33
Hawking, Stephen, 58, 170
heart, Pascal on, 164, 171
heavens, declare God's glory, 172
heliocentric system, 23–24, 34, 44, 94, 95, 150
Hinge of the World (play), 133
Hipparchus, 33, 95
Hoyle, Fred, 160
humility, 69, 73, 78, 131
Hurst, Lewis, 170–71
hypotheses, 75

ice glass, 146–47
image of God, 115
incarnation, 29, 41, 46, 61, 113, 135, 178
Inquisition, 24, 84, 89, 90, 98, 118–19, 130, 134, 135, 150, 158
intellectual discernment, 97
intuitive knowledge, 66
invisible world, 37
Isidore of Seville, 32n6

Jedin, Hubert, 111
Jerome, 68–69
John Paul II, Pope, 24–25, 28, 176
Jupiter, moons orbiting, 23, 35, 47, 68, 153

Kaiser, Walter C., Jr., 99–100
Keel, William, 105–6
Keeler, James Edward, 78
Kepler, Johannes, 32
King James Bible, 54
known knowns, 65
known unknowns, 65
Kostylo, Joanna, 145–46

Langford, Jerome, 44
laws of nature, as descriptive or purposeful, 74
Leeuwenhoek, Antonie van, 95
Leibniz, Gottfried Wilhelm, 95
Lemaître, Georges, 46
Leo X, Pope, 34
Leo XIII, Pope, 157
Letter to the Grand Duchess Christina of Tuscany (Galileo), 15–16, 27, 30, 31, 39, 149, 151–53, 179–213
Lewis, C. S., 40, 114–15
light, 58
Longland, John, 51
Lorini, Niccolò, 150

Ludwig, Emil, 63
Luther, Martin, 153

Maffeo, Sabino, 155–59
Maffi, Pietro, 158
Magritte, René, 101–2, 103
mankind, creation of, 66
material world, 37
mathematician vs. merchant, 76–77
McGrath, Alister and Joanna, 78
"Memorial" (Pascal), 164–66
Mentasti, Rosa Barovier, 144, 145–47
Mesnard, Jean, 15, 45–46, 164
Micanzio, Fulgenzio, 134
Michelson, A. A., 59–60
Milton, John, 81–82, 111, 119–20, 129, 134
miracles, 114–15
modern science, 73
mood of current age, 73
Moody, D. L., 169
More, Thomas, 51–52
Morrow, Lance, 103
moving earth, 149, 150
multiverse theory, 42, 75
Murano, 144–47

naturalism, 114–15
nature of truth, 43, 56, 66
Newcomb, Simon, 59, 127
Newton, John, 95, 107, 129
Nicene Creed, 55
Nicodemus, 168
"night of fire" (Pascal), 15, 45n10, 67, 82, 164–66
nonoverlapping magisteria (NOMA), 113
northern lights, 30

observational proof, 75
Offor, George, 51
Orr, H. Allen, 113

paradigm shifts, 68
Pascal, Blaise, 14, 15, 41, 45–46, 66, 95, 129, 163–66
perspective, 103
Peterson, Eugene, 172
Planck era, 124
Plantinga, Alvin, 114, 115
poetic language, in Scripture, 99–100
political correctness, 73
power of science, 123, 125
Ptolemy, 33

quantum theory, 60, 64

rationalism, limits of, 164
reason, and special revelation, 45
reductionism, 123
Rees, Martin, 58–59
Renaissance, 33
reticello glass, 147
Robertson, Alexander, 111
robes of a scientist, 176
Rowland, vicar of Great Wycombe, 51
Royal Astronomical Society, 170
Rumsfeld, Donald Henry, 65

Sandage, Allan, 132
Sarpi, Paolo, 110, 139–41, 142, 147–48
Schmidt, Brian P., 57n16
Schrödinger, Erwin, 42–43
science
    arrogance in, 78
    ascendancy of, 122–23
    driven by mood of our age, 39–40, 41
    as evolving discipline, 30
    faith in assumptions, 107
    as god, 16
    hidden agenda in, 34–35
    limitations of, 125, 132
    mysteries of, 77
    power-play agenda of, 120
    pride in, 69
    progress in, 35, 72, 124
    as reality, 56
    as set of partial truths, 30
    and torpedoing of careers, 174–75
    two faces of, 175, 176
    used to silence Scripture, 13
science and religion, separate vocabularies for, 103–4
scientific method, 56, 79–80
Scientific Revolution, 33, 45, 112, 117
scientism, 72, 82
Scripture. *See* Bible; book of Scripture
Sfondrati, Emilio, 150
Sidereus Nuncius (Galileo), 54, 141–42, 146
Siding Spring Observatory (Australia), 14
sin, 167
spiral galaxies, 57, 59, 78, 102, 124
spiritual discernment, 97
spyglass, 139, 141–42, 148
Spyker, John, 171–72
stars, formation of, 55
sunrise and sunset, 93–94

supernatural, 114
supernovae, 57n16

Tagore, Rabindranath, 45, 128, 130, 135
Teems, David, 50–51, 52, 57–58, 119
telescope, invention of, 139
theologians
    arguing beyond their trained discipline, 122
    arrogance among, 78
    flashing sword of power, 117–20
    misinterpretation of Genesis, 32
theological correctness, 73
theology
    as inferior and worthless discipline, 72
    as queen of sciences, 16
theory of general relativity, 60, 63
theory of special relativity, 60
Time Allocation Committee, 129
Tower of the Winds, 155, 156
Tozer, A. W., 169
"treachery of images," 101, 102
trees and wind, 35–37
truth, 132, 135
    of nature, 43, 56, 66
    never changes with time, 67
    never contradicts itself, 64–65
    physical and spiritual domains of, 13
    of Scripture and science, 24
Tsodilo hills (Botswana), 67
Tunstall, Cuthbert, 52
two books, 28, 35, 44, 47, 56. *See also* book of nature; book of Scripture
two cathedrals, 129
Tyndale, William, 49–54, 119

universe, finely tuned, 42, 74
unknown knowns, 65–66
unknown unknowns, 65
Urban VIII, Pope, 23, 52–53, 83, 86, 95, 96, 130, 133, 159

Vatican observatories, 128, 155–60
Venetian glass, 144–45
Viereck, George, 63
Voltaire, 111

Wolsey, Thomas, 52
Wootton, David, 111

Yates, Frances, 110

Zilio, Andrea, 147

Scripture Index

Genesis
book of 31, 32
1:3 58
26:15 177

Joshua
book of 204, 209, 210

Esther
book of 176
3:1 177
3:8–9 177
7:4 177
7:10 177

Job
book of 105
9:6 202
26:7 ....47, 201n2
27 201
38:4 47

Psalms
book of 99, 100
18:1 99
19 172
19:1 60
23:5 177
58:4–5 ....32n6
93:1 100
104:5 ....23, 100

Proverbs
8 ....133n15
8:26 ....132, 133

Isaiah
book of 174
45:3 173–74

Matthew
11:25–27 97
13:3–11 29
13:44 135
16:15–19 52

John
1 ....61, 67, 114, 131
1:14 136
1:14–18 67
3:1–8 168
3:8 ....42, 169
3:16 97
4:14 177
4:24 42
6:48 99
10:9 99
16:13 45
17:3 166
17:25 166
19:14 ....123n2

Romans
9:33 171

1 Corinthians
3:18–19 97
13:12 30

2 Corinthians
4:6 ....82, 131

Philippians
3:10–11 100

Hebrews
11:6 45

James
1:5 ....174n9